6th International Workshop on Multi-Rate Processes and Hysteresis

(MURPHYS2012)

Journal of Physics: Conference Series
Volume 585

Suceava, Romania
21 - 24 May 2012

ISBN: 978-1-5108-0076-2
ISSN: 1742-6588

Printed from e-media with permission by:

Curran Associates, Inc.
57 Morehouse Lane
Red Hook, NY 12571

CURRAN ASSOCIATES INC.
proceedings.com

Some format issues inherent in the e-media version may also appear in this print version.

Printed by Curran Associates, Inc. (2015)

Institute of Physics
Dirac House, Temple Back
Bristol BS1 6BE UK

Phone: 44 1 17 929 7481
Fax: 44 1 17 920 0979

techtracking@iop.org

Additional copies of this publication are available from:

Curran Associates, Inc.
57 Morehouse Lane
Red Hook, NY 12571 USA
Phone: 845-758-0400
Fax: 845-758-2634
Email: curran@proceedings.com
Web: www.proceedings.com

Table of contents

Volume 585

6th International Workshop on Multi-Rate Processes and Hysteresis (MURPHYS2012)
21–24 May 2012, Suceava, Romania

Accepted papers received: 14 January 2015
Published online: 25 February 2015

Preface

The International Workshop on Multi-Rate Processes and Hysteresis (MURPHYS) conference series focuses on multiple scale systems, singular perturbation problems, phase transitions and hysteresis phenomena occurring in physical, biological, chemical, economical, engineering and information systems. The 6th edition was hosted by Stefan cel Mare University in the city of Suceava located in the beautiful multicultural land of Bukovina, Romania, from May 21 to 24, 2012. This continued the series of biennial multidisciplinary conferences organized in Cork, Ireland from 2002 to 2008 and in Pécs, Hungary in 2010.

The MURPHYS 2012 Workshop brought together more than 50 researchers in hysteresis and multi-scale phenomena from the United State of America, the United Kingdom, France, Germany, Italy, Ireland, Czech Republic, Hungary, Greece, Ukraine, and Romania. Participants shared and discussed new developments of analytical techniques and numerical methods along with a variety of their applications in various areas, including material sciences, electrical and electronics engineering, mechanical engineering and civil structures, biological and eco-systems, economics and finance.

The Workshop was sponsored by the European Social Fund through Sectoral Operational Program Human Resources 2007-2013 (PRO-DOCT) and Stefan cel Mare University, Suceava. The Organizing Committee was co-chaired by Mihai Dimian from Stefan cel Mare University, Suceava (Romania), Amalia Ivanyi from the University of Pecs (Hungary), and Dmitrii Rachinskii from the University College Cork (Ireland).

All papers published in this volume of *Journal of Physics: Conference Series* have been peer reviewed through processes administered by the Editors. Reviews were conducted by expert referees to the professional and scientific standards expected of a proceedings journal published by IOP Publishing. The Guest Editors wish to place on record their sincere gratitude to Miss Sarah Toms for the assistance she provided during the publication process.

More information about the Workshop can be found at <http://www.murphys.usv.ro/>

Mihai Dimian and Dmitrii Rachinskii

Guest Editors for *Journal of Physics: Conference Series*

Proceedings of the 6th International Workshop on Multi-Rate Processes and Hysteresis

Workshop Co-organizers:

USU Universitatea Ştefan cel Mare Suceava

UCC Coláiste na hOllscoile Corcaigh, Éire University College Cork, Ireland

PÉCSI TUDOMÁNYEGYETEM UNIVERSITY OF PÉCS

MURPHYS May 21st - 24th, 2012 Suceava, Romania | 6th International Workshop on MULTI-RATE PROCESSES & HYSTERESIS in Mathematics, Physics, Engineering and Information Sciences

International Steering Committee:

Petru ANDREI	*Florida State University (USA)*
Zalaman I BALANOV	*Netanya Academic College (Israel)*
Martin BROKATE	*Technical University Munich (Germany)*
Morten BRØNS	*Technical University of Denmark (Denmark)*
Rod CROSS	*University of Strathclyde (UK)*
Karin DAHMEN	*University of Illinois, Urbana-Champaign (USA)*
Mihai DIMIAN (Co-Chairman)	*Stefan cel Mare University (Romania)*
Dietrich FLOCKERZI	*Max Planck Institute, Magdeburg (Germany)*
Gary FRIEDMAN	*Drexel University (USA)*
Amalia IVANYI (Co-Chairman)	*University of Pecs (Hungary)*
Leonid KALACHEV	*University of Montana (USA)*
Olaf KLEIN	*Weierstrass Institute (Germany)*
Pavel KREJCI	*Academy of Sciences (Czech Republic)*
Jorge LINARES	*University of Versailles St. Quentin (France)*
Jeff MOEHLIS	*University of California, Santa Barbara (USA)*
Robert E O'MALLEY, Jr	*University of Washington (USA)*
Anatoly NEISHTADT	*Academy of Science (Russia)*
Nikola POPOVIC	*University of Edinburgh (UK)*
Dmitrii RACHINSKII (Co-Chairman)	*University College Cork (Ireland)*
Sergei S SAZHIN	*University of Brighton (UK)*
Alexandru STANCU	*Al. I Cuza University, Iasi (Romania)*
Vladimir SOBOLEV	*Samara Sate University (Russia)*
Ciro VISONE	*University of Sannio (Italy)*

Journal of Physics: Conference Series **585** (2015) 011002 doi:10.1088/1742-6596/585/1/011002

Peer review statement

All papers published in this volume of *Journal of Physics: Conference Series* have been peer reviewed through processes administered by the proceedings Editors. Reviews were conducted by expert referees to the professional and scientific standards expected of a proceedings journal published by IOP Publishing.

Vector Preisach modeling of magnetic materials under stress

A Ktena[1]
Department of Electrical Engineering, *TEI of Chalkida*,
Psahna, Evia 34400, GREECE

E-mail: aktena@teihal.gr

Abstract. The Preisach formalism is used to model magnetic hysteresis loops in soft magnetic materials subject to tensile stress. The model uses the Stoner-Wohlfarth mechanism of coherent rotation and dispersion of easy axes to capture the vector response of the magnetization. The Preisach density is constructed as the weighed sum of normal probability density functions (pdf) for the regions of high and low induction. The model parameters reflect the effect of strain: increased pinning, modelled by the central pdf parameters; enhanced anisotropy dispersion modelled by the angular dispersion of easy axes. Upon removal of the tensile stress, compressive residual stresses give rise to effective demagnetizing fields leading to lower differential permeability with a two-peak profile. As deformation levels increase, the amplitude of and the relative distance between the two permeability peaks changes which is reflected in the side density parameters. Modelling results are in qualitative agreement with the experimental data. The potential and limitations of the model are discussed.

1. Introduction

Magnetization is coupled to strain via the magnetoelastic energy of the material. The resulting property is magnetostriction and is a measure of the deformation induced by a magnetic field as well as of the propensity of a material to deform under an applied field. Magnetization processes are influenced by imperfections and material defects such as voids and impurities (0D), dislocations (1D), grain boundaries and interfaces (2D) and of course cracks or pores (3D). 0D imperfections cause a localized variation of the anistotropy profile and an imbalance in the stress tensor around the point defect. 1D defects are abrupt changes in the regular ordering of atoms along the dislocation line while 2D defects are the result of the variation of the stress tensors at the interfaces and grain boundaries. The number of dislocations per unit volume is called dislocation density and they are measured per unit area in a plane. As dislocations propagate they may generate more dislocations assisted by other defects, grain boundaries, and surface irregularities.

The dislocation density increases and shifts with stress, affecting the magnetic response of a material. The effect of dislocations on the magnetic properties [1] of crystals has been studied both theoretically and experimentally. The results reveal a relationship between stresses, applied or residual, and macroscopic magnetic properties, such as the coercivity, the differential susceptibility, the remanence, the shape and the area of minor and major hysteresis loops, as well as microscopic processes such as domain wall mobility [2-3].

[1] To whom any correspondence should be addressed.

In a rigid wall approximation model for domain wall motion [2] the inverse of the initial permeability and the coercive field are found to be proportional to the square root of the dislocation density. Experimental studies have shown that the motion of dislocations is more impeded by the 90° walls, also related to magnetostriction, than the 180° walls [4].

From the point of view of mechanical properties, stresses have been found to give rise to long range internal stresses, originating at the intergranular and intragranular level. The intergranular component is related to the grain-boundary structure and the incompatibility between grains, which can be considered as a soft mechanical region, and the high density of dislocations acting as hard mechanical region. As plastic strain increases, dislocations form tangles and forests of high density and compete against regions of lower dislocation density leading to intragranular internal stresses. A material in the plastic deformation region can be modeled as consisting small volumes under tensile stresses (hard regions) separated by larger volumes under compressive stresses (soft regions) [5]. These long range internal stresses, affect the magnetoelastic energy and the magnetization processes in the strained material.

Stress-dependent magnetization has been modeled at the atomic [6], the mesoscopic [2, 7-8] and the macroscopic level [9-11]. The Jiles-Atherton model [9] and the moving Preisach model [10-11] have established that the parameters of the characteristic Preisach density are related to microscopic parameters such as the dislocation density and the grain size. However, these models are scalar failing fail to capture the stress – related magnetization rotation and they do not take into account the effect of internal stresses. The model presented in [12] is a vector model based on the Preisach formalism which reproduces successfully the magnetization and permeability dependence on the applied field as a function of stress and whose properties and limitations are discussed in this work.

2. Preisach modeling

The Preisach formalism [13] treats the magnetization response of a magnet exposed to an externally applied field as the superposition of the responses of elementary hysteresis operators. The novelty of the model is that, departing from the classical thermodynamic treatment of hysteresis, it is based on the phenomenology of hysteresis and not on the physical properties of the material being studied.
Hysteresis in magnetic materials is a highly non-linear phenomenon, almost synonymous to magnetism, the origins of which lie in the atomic structure of a material and the ensuing complex network of short- and long-range interactions. The etymology of the word suggests that in a system exhibiting hysteresis (means delay in Greek) the output, y, lags the input, x. Furthermore, the present output is a complicated non-linear function of the present input as well as of the past input values. From a stability point of view, in a system exhibiting hysteresis, for a given input value, there may be more than one possible equilibrium states depending on previous equilibrium states, *ie*, the history of the system. Therefore, a system with hysteresis is also a system with memory.

According to the Preisach formalism, the elementary hysteresis operator, γ_{ab} (figure 1) is a rectangular loop with values *+1* and *-1* bounded by two critical fields, a and b, with $a > b$, also known as the upper and lower switching fields, respectively, because the magnetization 'switches' to the *+1* – state (or *-1* – state) for $h>a$ (or $h<b$).

One of the nicest features of the Preisach formalism is the elegant inclusion of interactions: if there are no interactions, the upper and lower fields are equal, $a=b$; in the presence of interactions it becomes harder to switch to the *+1* – state ($|a| > |b|$) or the -1 – state ($|a| < |b|$) depending on the direction of interactions (figure 1). The shift of the elementary loop is then equivalent to the interaction field $h_i = ((a+b)/2)$ while the half-width of the elementary hysteresis loop $h_c = ((a-b)/2)$ can be thought of as the analogous to the coercive field of the loop.

At this point, it is important to note that the elementary hysteresis operator γ_{ab} does not represent a specific, physical entity of the material, such as a grain, a particle or a domain; it is a statistical variable representing the fields necessary to cause the switching of the magnetization of entities with identical coercivities and experiencing the same amount of interactions. The values allowed for a and

b define the Preisach plane (figure 1) over which a probability density function (pdf) is constructed, $\rho(a,b)$.The value of $\rho(a,b)$ for a given point (a,b) is a weight representing the probability that a physical entity in the material is characterized by the given upper and lower switching fields. $\rho(a,b)$ can be inferred from minor loop measurements and is the footprint of the hysteresis processes inside a material.

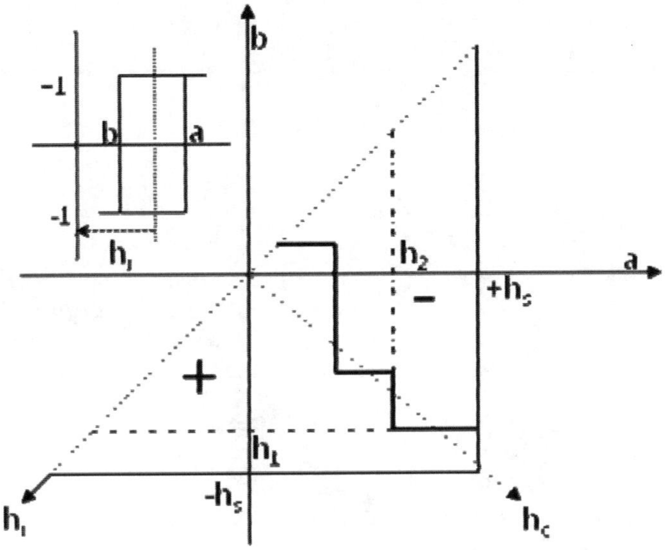

Figure 1. Classical Preisach Model: the elementary hysteresis operator with upper and lower switching fields a and b (top left); the triangular Preisach plane, over which the density $\rho(a,b)$ is defined, bounded by a=+h_s, b=-h_s and the interactions axis h_i; the staircase boundary separating regions of positive and negative magnetization for an alternating field sequence $+h_s$, h_1, h_2, h_3, h_4 with $h_1,h_3<0$, $h_2,h_4>0$.

The magnetization $M(H)$ of the material for a given applied field H is then computed by:

$$M(H) = \iint_{a>b} \rho(a,b)\gamma_{ab} H da db \qquad (1)$$

The Classical Preisach Model (CPM) described in equation 1 is an inherently scalar model accounting only for irreversible processes (switching) and has performed very well over the years for a number of systems and materials for which the following assumptions are valid or allowed: 1) a field H results in the same change of magnetization ΔM regardless of the existing magnetization configuration, *eg* whether the field is applied at the remanent or the demagnetized state 2) the magnetization processes are one-dimensional (1D), confined along the axis of the applied field 3) reversible processes are decoupled from irreversible processes and can be added on later to the response computed with equation 1. As a result of the above assumptions, CPM i) fails to reproduce non-congruent minor loops, *ie* minor loops which depend on the magnetization configuration as well as on the applied field ii) cannot account for the off-easy axis contributions of the magnetization in non-perfectly oriented media iii) fails to capture the effect of magnetization rotation leading to reversible as well as irreversible processes.

To relax the first assumption, the moving model [14] has been proposed which, along the lines of the mean-field theory, introduces the idea of an effective field, H_{eff}, rather than the applied field H, being responsible for the state of γ_{ab}:

$$H_{eff} = H + \alpha M \qquad (2)$$

where α is a constant or a function of the magnetization configuration M on which H is applied. The moving model resulted in more realistic loops, levying the congruency.

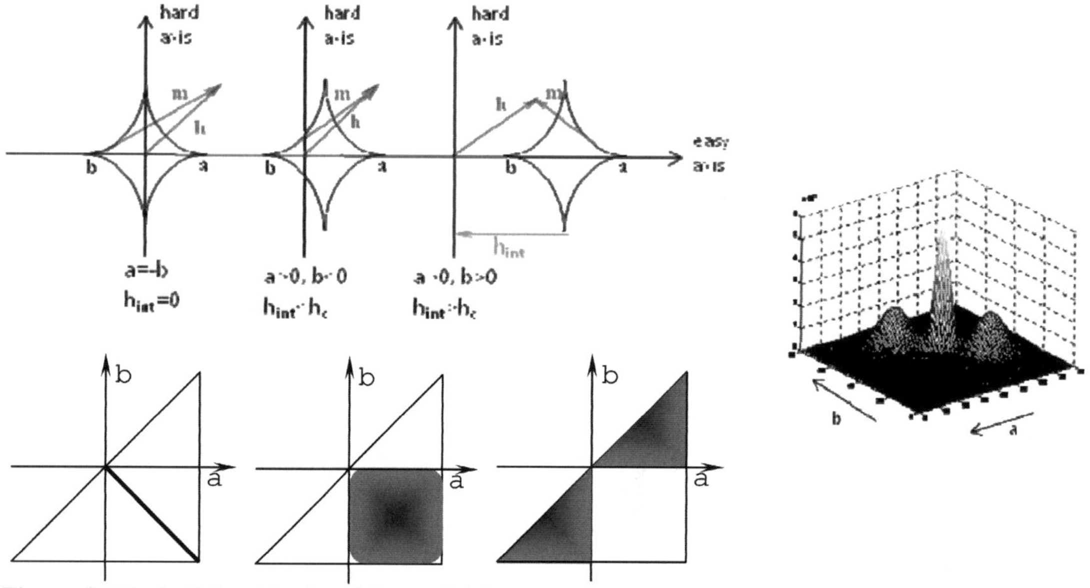

Figure 2. The building blocks of the model for stress dependent magnetization: the SW hysteresis operator accounting for interactions, the probability density function and the Preisach plane

The vector hysteresis model presented in [15] has been developed along the lines of the Preisach formalism. The scalar elementary hysteresis operator γ_{ab} is replaced by a vector one, such as the Stoner-Wohlfarth mechanism of coherent rotation of the magnetization for an isolated, single domain, ellipsoidal particle with easy axis, x, along the long axis of the ellipse, and a transverse hard axis, y. The mechanism is conveniently depicted by the, so called, SW – astroid (figure 2) which is the locus of solutions to the equation

$$h_x^{2/3} + h_y^{2/3} = 1 \qquad (3)$$

where h_x and h_y are the components of the input field H, normalized to the particle's anisotropy field H_k, along the easy and hard axis, respectively. When a field $\boldsymbol{h}$ is applied, the angle ϕ of the magnetization vector, $\boldsymbol{m}$, to the easy axis, is the solution to the transcendental equation

$$h_x \, tan\,\phi - h_y + sin\,\phi = 0 \qquad (4)$$

which is the result of the minimization of the free energy equation for an isolated, ellipsoidal magnetic particle with uniaxial anisotropy under the influence of an applied field, h.

Interactions are included along the lines of the Preisach formalism, *ie* the SW operator, γ_{ab}, is allowed to shift along the easy axis. Then, the magnetization vector response $\boldsymbol{M}$ for an applied field $\boldsymbol{H}$ is given by

$$\mathbf{M}(\mathbf{H}) = \iint_{a>b} \rho(a,b) \gamma_{ab} \mathbf{H} \, da\, db \qquad (5)$$

Equation 5 describes the response for a system where the magnetic domains share the same easy axis, *ie* a perfectly oriented system with uniaxial anisotropy. For a more realistic response, dispersion of orientations is added by superimposing the responses of angularly distributed perfectly oriented models:

$$\mathbf{M}(\mathbf{H}) = \int \rho(\theta) d\theta \iint_{a>b} \rho(a,b) \gamma_{ab} \mathbf{H} da db \tag{6}$$

where $\rho(\theta)$ is a probability density function of angles of anisotropy axes.

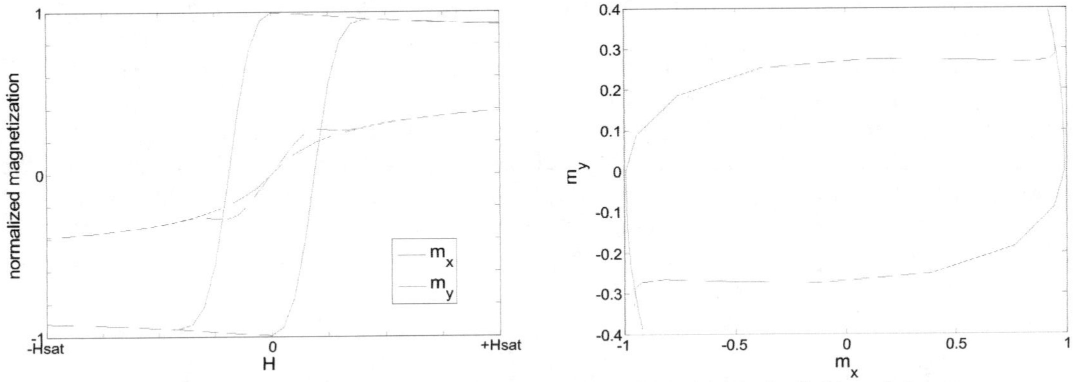

Figure 3. Computed hysteresis loops along the easy (x) and hard (y) axis (left) and the transverse m_y vs the longitudinal m_x component of the magnetization (right)

Figure 3 demonstrates the vector properties of the model depicting the dependence of the magnetization along the easy and the hard axis for a major loop field sequence applied at a $\pi/6$ angle to the easy axis. The computed results are in agreement with experiments [15] and the loops predicted by this model are non-congruent [16-17].

3. Stress-dependent magnetization and Preisach modeling

Researchers from several laboratories have reported results on the effect of stress on material parameters such as the Barkhausen Noise, the coercivity, the remanence or the permeability of ferromagnets [4,18-19]. The results concern measurements in loaded and unloaded states. In the loaded case, the applied stress has the effect of an effective field assisting the magnetization processes along the direction of application. Barkhausen Noise, remanence and loop squareness increase with strain as the material enters the plastic deformation region but they decrease as the material approaches fracture. The increase is associated with the increase of dislocations but as the material approaches fracture, further domain wall movement is progressively hindered and rotations are more prominent.

Upon unloading, the Barkhausen Noise is lower with respect to the loaded case. As described in the introduction, the deformed material may be considered as consisting of a hard mechanical phase with high dislocation density and a soft phase with low dislocation density resulting in long range internal stresses, mostly compressive in nature, which play a predominant role upon the removal of the applied stress [18-19]. Remanence is found to decrease with strain which is an indication of the presence of transverse domains. Domain rotation and and non-180° wall motion become dominant as magnetization reversal mechanisms [19] while the double-peaked Barkhausen Noise envelop and permeability, characteristic of strained materials [figure 4], appears when the magnetization is perpendicular to the easy magnetization axis and is associated to 90° domain wall rotation [18].

Based on the above results, it is clear that a macroscopic hysteresis model should be able 1) to incorporate the effect of long-range stress-related interactions and 2) to account for the rotation of transverse domains 3) to account for the effect of compressive fields.

The vector model proposed for the modelling of stress dependent magnetization [12] is based on equation 6 but uses a weighed mixture of gaussians (figure 2) to construct the Preisach density, *ie* a central density, ρ_2, and two side densities, $\rho_{1,3}$ [17]:

$$\rho(a,b) = w_1\rho_1(a,b) + w_2\rho_2(a,b) + w_3\rho_3(a,b) \tag{7}$$

where $w_1+w_2+w_3=1$.

The central density, ρ_2, is the pdf of operators with $a>0,\ b<0$ (figure 2), accounts for the switching occurring around the coercivity and corresponds to the second and main peak in the permeability curve. The first side density, ρ_1, is the pdf for operators with $a,b > 0$, *ie* the interaction field is greater than the coercivity and opposite to the direction of the applied field. It accounts for the first peak observed in the differential permeability (figure 4), when the field is decreasing from positive saturation, before remanence. This peak may be related to the switching that occurs under the effect of strong stress-related magnetostatic interactions, before the applied field becomes zero. The third side density, ρ_3, has an odd symmetry to the first one, with $a,b<0$, *ie* the interaction field is greater than the coercivity and opposite to the direction of the applied field. It is responsible for the third peak observed in the permeability before reaching saturation (descending curve). Approaching negative saturation, there are magnetic domains which experience positive magnetostatic interactions high enough to oppose their alignment with the field.

4. Results

The model described by equations 6 and 7 has been used to calculate hysteresis loops and the corresponding differential permeability curves as well as the derivative of the latter ones, for various sets of model parameters in order to reproduce the effect of plastic strain on: the maximum inductance, B_{max}; the remanence, B_r; the coercivity, H_c; the value and the location of the primary and secondary peak of the differential permeability. Note that B_{max} is used instead of saturation magnetization, for practical purposes, in order to signify the magnetization value beyond which no further irreversible processes occur. Following the *ceteris paribus* principle, the role of the following parameters is examined: central distribution parameters, μ_2 and σ_2, side distribution parameters, $\mu_{1,3}$ and $\sigma_{1,3}$, weights $w_{1,2,3}$.

The results are compared qualitatively against experimental data obtained on commercial low carbon electrical steel laminations of 0.5mm thickness. Tensile stress has been applied at constant strain rates, namely 0.1, 0.5 and 1 mm/min, and the hysteresis loop of the sample was measured at various strain levels from 0 till fracture while the sample was loaded and after unloading. The hysteresis loop measurements were obtained using the in-house ac hysteresiograph at various frequencies from 0.1 to 1 Hz. The results are in arbitrary units (a.u.) because the hysteresiograph is not calibrated. The magnetic field and the tensile stress are applied along the length of the lamination; the output is measured along the same direction. The output of the hysteresiograph is a voltage pulse proportional to the differential permeability, μ_{diff}, of the sample, the integration of which yields the $B(H)$ loop. The results presented in this work concern hysteresis loops at fields high enough to achieve B_{max}; frequency of 0.1 Hz; strain rate of 0.5 mm/min; samples after unloading, in order to demonstrate and discuss the performance of the model in the unloaded case, where the effect of residual stresses is prominent.

Fig. 4 shows the effect of strain on the magnetization process. At zero strain, the hysteresis loop is more square and the maximum value of the permeability curve occurs near the coercivity, as expected. As the strain increases, the maximum inductance B_{max} and the remanence B_r decrease, the coercivity H_c increases, the maximum permeability μ decreases and the permeability curve has two peaks: a primary peak around the coercivity and a secondary one as remenance is approached from saturation: switching occurs even before the field changes sign and the permeability changes in a non-monotonic way. This non-monotonic field dependence of the permeability, often referred to as negative permeability or 'dip' in permeability [6-7,11,21], is associated with residual stresses and will be

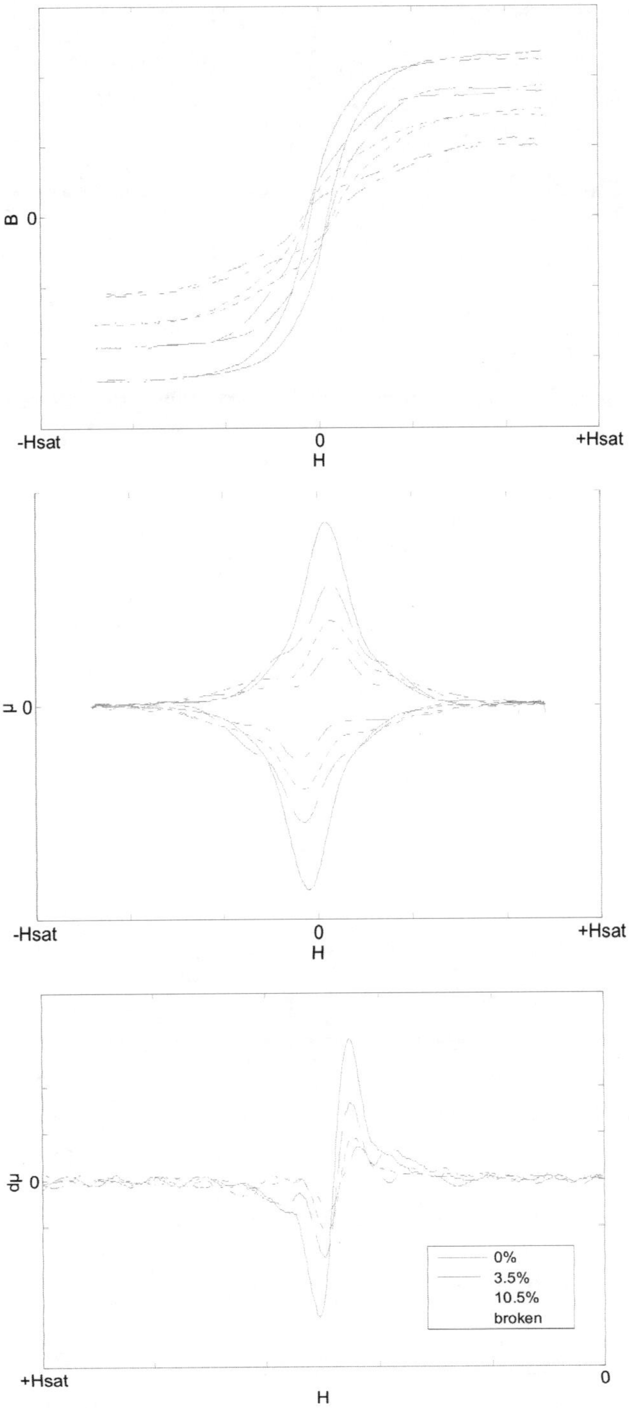

Figure 4. From top to bottom: Measured hysteresis loop *B(H)*; differential permeability, μ; derivative of the differential permeability measurement, $d\mu$, for the descending curve at various strain levels.

discussed in the following section. A useful tool in evaluating the effect of strain on permeability is its derivative $d\mu$ (figure 4).

Figure 5 depicts the calculated hysteresis and permeability curves for various values of angular dispersion ϕ while keeping all other model parameters constant. This parameter takes into account the effect of easy axes dispersion which is expected to increase with strain. As the angular dispersion increases the maximum inductance B_{max} and the remanence B_r as well as the maximum permeability μ decrease which is in agreement with the experimental evidence. However, the coercivity of the computed loops decreases which indicates that angular dispersion is not sufficient to explain the effect of strain on the magnetization process.

Next, the effect of the mean values of the side densities, μ_1 and μ_3, is examined. Figure 5 shows the calculated results for side densities centered at +/- $0.1\mu_2$, $0.2\mu_2$, $0.3\mu_2$ from the central one, μ_2,. B_{max} and H_c are practically unaffected but the squareness of the hysteresis loop at the remanence and the coercivity as well as μ decrease as the side densities move away from the central density to stronger fields. Furthermore, the secondary peaks of the permeability are enhanced and occur at stronger fields. This is consistent with the effect of increasing strain.

Varying the variance of the side densities, it is possible to fine tune the peak heights in the permeability curve: narrower distributions lead to enhanced secondary peaks with respect to the primary one. Finally, the weights of the densities control the relative heights between the primary and secondary permeability peaks.

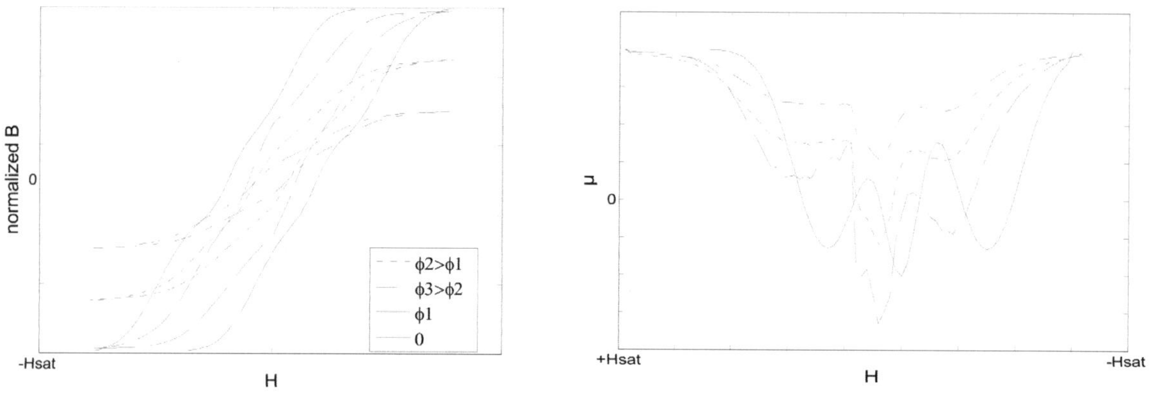

Figure 5. Calculated hysteresis loops (left) and permeability curves for the descending branch (right) at angular dispersions $0<\phi_1<\phi_2<\phi_3$.

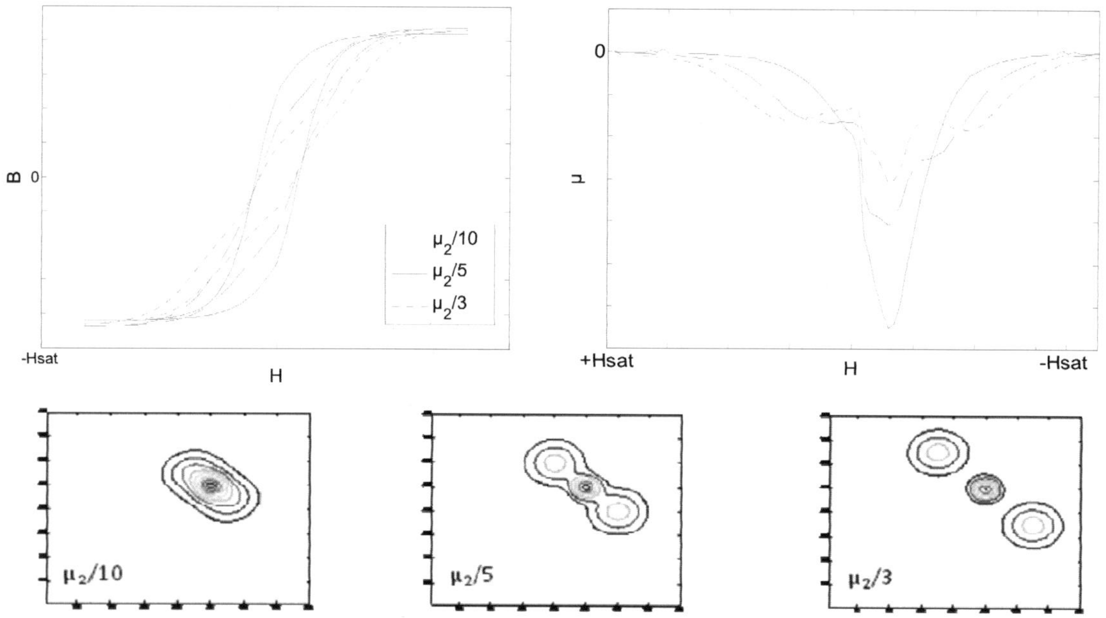

Figure 6. Calculated hysteresis loops (left) and permeability curves for the descending branch (right) for side densities located at +/- $0.1\mu_2$, $0.2\mu_2$, $0.3\mu_2$ with respect to the central density; the respective contour plots of the characteristic density $\rho(a,b)$ are shown at the bottom.

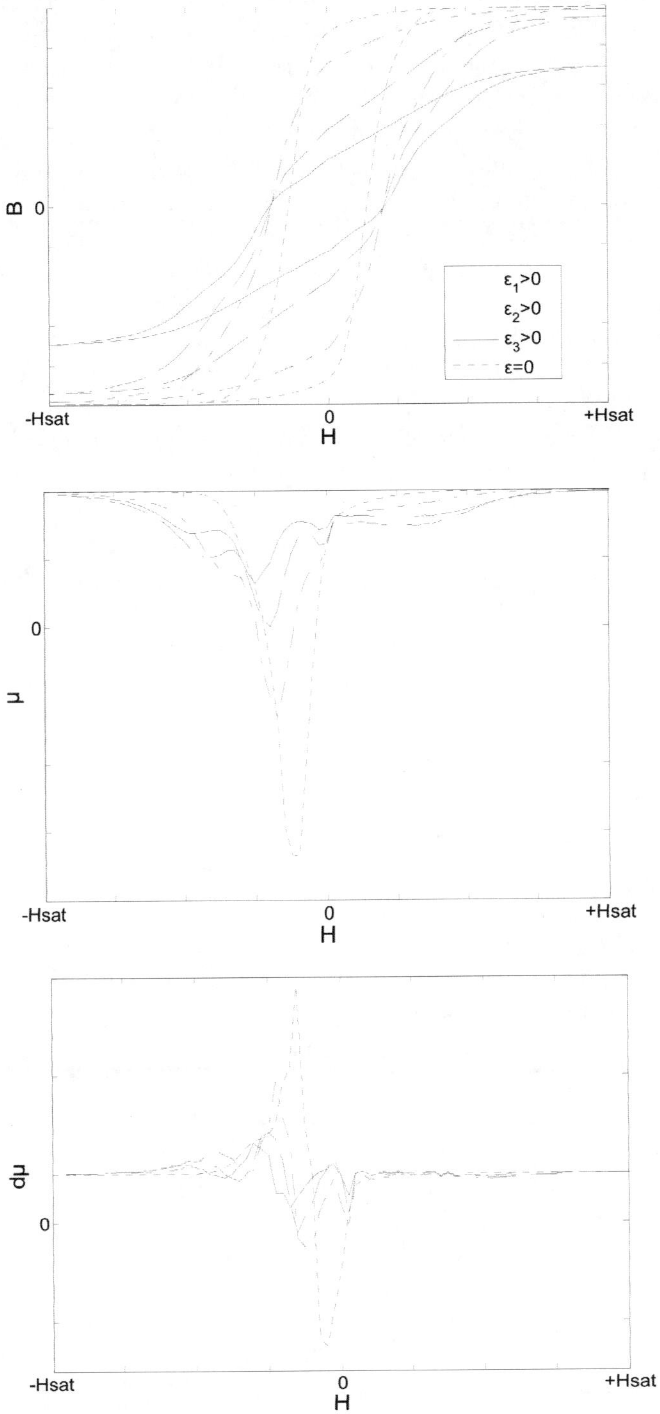

Figure 7. From top to bottom: Calculated hysteresis loop *B(H)*; differential permeability, μ; derivative of the differential permeability measurement, $d\mu$, for the descending curve at various strain levels.

The results presented so far concern the effect of each model parameter while the rest are kept constant. Figure 7 presents the results obtained when all parameters are varied in order to reproduce the phenomenology observed experimentally with increasing strain (figure 4). At zero strain, the effect of the side densities is negligible and the angular dispersion is lowest. In order to generate the curves for increasing strain levels, $\varepsilon_1<\varepsilon_2<\varepsilon_3$, the mean value of the central density μ_2 increases, μ_1 and μ_3, move away from μ_2 towards higher fields, the ratio w_1/w_2 increases and so does the angular dispersion.

5. Discussion

The parameters of the central distribution, μ_2 and σ_2, control the coercivity and the squareness at the coercivity respectively and as a consequence they control the value and location of the primary peak of the differential permeability. Therefore, we can claim that the mean μ_2 is related to the switching via 180° domain wall movement which is associated to pinning. As strain increases, the dislocation density increases, grains become smaller and as a result pinning increases. The parameter σ_2 is related to the mixture of exchange and magnetostatic interactions, *ie* short-range vs long-range interactions in the material.

The effect of angular dispersion is very strong (figure 5) affecting the maximum induction, the loop squareness parameters, the coercivity, and the values of the primary and secondary permeability peaks.

All of the above decrease with increasing dispersion which underlines the important role of rotations. The increase in strain and the build-up of residual stresses leads to smaller grains with different crystalline anisotropy axes. The applied magnetic field forces the domain magnetizations to rotate towards its direction in order to minimize their energy. This leads to lower induction levels and lower squareness at the remanence and coercivity.

The side densities represent the effect of the build-up of the internal demagnetizing field due to residual stress. As residual stresses increase, the magnetoelastic energy, which is a function of the magnetostriction and the elastic properties of the material, increases. To counterbalance this increase in energy, an effective mean field opposing the applied field is established. Increased mean values, μ_1 and μ_3, result in lower coercivity, remanence and permeability peak value. The secondary permeability peaks are enhanced and shift away from the zero field as these parameters increase suggesting that the internal demagnetizing fields are stronger are strain increases.

Finally, the weights, $w_{1,2,3}$, of the three pdfs are related to the degree magnetic domains are affected by internal demagnetizing fields and the strength of these fields with respect to their coercive fields and other interactions they experience.

Based on the above analysis, the mechanisms and factors affecting the stress-dependent magnetization processes are three: i) increased pinning ii) dispersion of anisotropy axes iii) internal demagnetizing fields. The first is more important at low strain levels when dislocations increase and propagate fairly easily. This explains why at low strain levels the most prominent increase in coercivity is observed [4]. As strain increases and the material is well into the plastic deformation region, the other two mechanisms prevail leading eventually to deterioration of magnetic structure and material failure. These observations hold for both loaded and unloaded materials but the effect of residual stresses is more obvious in the later case.

Micromagnetic calculations using the Landau-Lifshitz equation to study the effect of dislocations and the orientation of the crystallographic axes on the magnetization have shown that magnetoelastic energy becomes more important as dislocations increase enforcing the magnetic dipoles to align along preferable directions corroborating our results [8]. Along the same lines, RFIM modeling has shown the effect of an internal demagnetizing field due to long-range dipolar interactions [6] aiming to minimize the energy of the material. These interactions are enhanced in high inductance regions by the residual compressive stresses causing rotation.

The model presented reproduces the double peak observed in the permeability the origins of which are still under investigation by several researchers. The interest is focused on the finding that the permeability becomes negative approaching remanence. Measurements of the Preisach density on samples deformed under tensile stress, designed to be used with a moving Preisach model [14] have associated the dip in susceptibility with negative pdf values. Negative pdf values have also been measured in the FORC analysis [20]. Monte Carlo simulations have shown that the negative susceptibility around zero field is indeed associated to dislocations because of spin fluctuations [7]. Finally, negative Barkhausen jumps may be possible if the increase in the local Zeeman energy is associated with a local magnetic configuration with the magnetization opposing the applied field [21]. In our approach, the phenomenology is reproduced using a weighed sum of densities where the central density represents the coercivity and magnetostatic or exchange interactions of the material and the side density accounts for the effect of the internal demagnetizing fields established with increasing strain.

The identification of this model is still under investigation. Preliminary results show that B_{max} is controlled by angular dispersion, H_c by the central density ρ_2. The identification of the other parameters should be based on the differential permeability curve rather than the hysteresis curve because the side densities determine the position and relative heights of the permeability peaks. The derivative of the differential permeability is a useful metric in that direction since it is related to the location of the permeability 'dip'.

6. Conclusion

A vector model utilizing the Stoner-Wohlfarth mechanism of coherent rotation and the statistical approach to interactions of the Classical Preisach Model is proposed for the modelling of the hysteresis phenomenology in stress dependent magnetization. The effect of residual stresses is modelled by a weighed mixture of pdfs allowing for nonzero density values in the triangular areas of the Preisach plane where the interaction fields are stronger than the coercive fields on the underlying assumption that the residual, mostly compressive, stresses establish a long-range demagnetizing field. The modelling results are in qualitative agreement with experimental findings corroborating the results of other modelling approaches. An interesting remark is that the negative permeability observed in strained materials approaching the remanent state of magnetization is controlled by the side densities which represent the long-range demagnetizing fields. The model parameters are related to material parameters but the identification should be based on the differential permeability curve along with its derivative as well as on the hysteresis loop.

7. References

[1] Sato A, Nagao M 1992 *Magnetic properties of dislocations, in Concise encyclopedia of magnetic and superconducting materials,* Ed. K.H.J. Buschow (Elsevier)
[2] Seeger A, Kronmüller H, Rieger H, and Träuble H 1964 *J. Appl. Phys.* **35**, 740
[3] Takahashi S, Zhang L and Ueda T 2003 *J. Phys.: Condens. Matter* **15** 7997–8002
[4] Iordache V E, Hug E, Buiron N 2003 *Mat. Sci. Eng.* **A359** 62-74
[5] Mughrabi H 1983 *Acta Metall. Mater.* **31** 1367
[6] Perez-Benitez J A, Capo-Sanchez J, Padovese L R 2009 *Comp. Mat. Sci.* **44** 850–857
[7] Yamaguchi K, Tanaka S, Nittono O, Yamada K, Takagi T 2006 *Physica B* **372** 251–255
[8] Van de Wiele B, Dupre L, Olyslager F 2008 *Physica B* **403** 342–345
[9] Sablik M J 2001 *J. Appl. Phys.* **89** 5610–3
[10] Dupre L, Sablik M J, Van Keer R and Melkebeek J *2002 J. Phys. D: Appl. Phys.* **35** 2086–2090
[11] Perevertov O 2003 *J. Phys. D: Appl. Phys.* **36** 785–790
[12] Ktena A, Hristoforou E 2012 *IEEE Trans Mag* **48**(4), 1433-1436
[13] Preisach F 1935 *Zeitschrift für Physik,* **94**, 277-301
[14] Oti J and Della Torre E 1990 *J. Appl. Phys.* **67** 5364-5366
[15] Ktena A and Charap S H 1993 *IEEE Trans. Magn.* **29** (6) 3661-3663
[16] Ktena A, Fotiadis D I, Spanos P D, Berger A and Massalas C V 2002 *Int. J. Eng. Sci.* **40** (20) 2235-2247
[17] Ktena A, Fotiadis D I, Berger A and Massalas C V, 2004 *Physica B* **343** 101-106
[18] Stupakov O, Uchimoto T and Takagi T 2010 *J. Phys. D*: Appl. Phys. **43** 195003
[19] Landgraf F J G, Emura M, Teixeira J C, de Campos M F 2000 J. *Magn. Magn. Mat.* **215-216** 97-99
[20] Roberts A P, Pike C R and Verosub K L 2000 *J. Geo. Res.* **105**(B12) 28461-28475
[21] Yang S, Beach G S D, and Erskine J L 2006 *J. Appl. Phys.* **100** 113914

Journal of Physics: Conference Series **585** (2015) 012002 doi:10.1088/1742-6596/585/1/012002

Scalar and vector hysteresis simulations using HysterSoft

M Dimian[1] and P Andrei[2]
[1]Department of Computers, Electronics and Automation, *Stefan cel Mare University*, Suceava 720229, Romania
[2]Department of Electrical and Computer Engineering, *Florida State University,* FL 32310, USA

E-mail: pandrei@fsu.edu

Abstract. Hysteresis modeling has become an important research area with many applications in science and engineering. In this article we present a unified and robust simulation framework designed to perform scalar and vector hysteresis modeling. The framework is based on HysterSoft©, which is a simulation platform that can be interfaced with other libraries and simulation programs to model various aspects of hysteresis. We describe the main features of our simulation framework by focusing on scalar and vector hysteresis modeling, direct and inverse modeling, dynamic hysteresis modeling, first-order reversal-curves analysis, identification of the scalar and vector Preisach distribution function using an experimental first-order reversal-curves, noise passage analysis through hysteretic systems, and thermal relaxation in scalar and vector hysteresis. The simulation modules, the user-defined features, and various parameter identification techniques are also presented.

1. Introduction

Scalar and vector hysteresis can be found in many areas of engineering, physics, material science, biology, economics, and social sciences. Due to the wide spectrum of areas in which hysteresis is observed and to the fact that the origins of hysteresis are often multiple and unclear, there exist a large number of theoretical models for hysteresis in the literature. In this article we present, HysterSoft© - a robust simulation environment designed to perform scalar and vector hysteresis modeling [1]. The software was developed at Florida State University (Tallahassee, USA) in collaboration with research groups from other institutions worldwide: the Institute of Industrial Electronics and Material Science at Vienna Institute of Technology (Vienna, Austria), the Department of Computers, Electronics and Automation at Stefan cel Mare University (Suceava, Romania), and Department of Physics at Alexandru Ioan Cuza University (Iasi, Romania).

One of the most important features of HysterSoft© is its universality in the sense that it does not make reference to particular hysteretic systems such as magnetic or mechanic hysteresis, but focuses on the mathematical modeling of hysteresis in general. For this reason HysterSoft© can be used for both educational and research purposes and in a variety of disciplines and its applications may involve finite elements or other physics-based simulations. HysterSoft© is currently employed as the main editor for ferromagnetic and ferroelectric materials in RandFlux©, which is a device and circuit simulator developed at Florida State University [2]. Models defined in HysterSoft© can be easily imported in RandFlux© and used to simulate electronic circuits containing hysteretic inductors, transformers, and capacitors. In addition, HysterSoft© can also be used to compute scalar and vector first order reversal curves diagrams, identify the model parameters from experimental data, conduct temperature and stress dependent simulations, perform noise passage analysis in hysteretic systems, etc. Most of these simulations can be performed using any hysteretic model either predefined or user-defined. The program can also be used as a dynamic link library and called from other programs such as Matlab, Simulink, Java, or C++.

A list of the most important problems that can be addressed by using HysterSoft© is given below:

- Scalar and vector hysteretic processes as well as inverse hysteretic problems;
- Dynamic hysteresis phenomena such as rate or frequency dependent hysteretic processes;
- Computation of first order reversal curves (FORCs) and FORC diagrams;
- Identification of the Preisach distribution function using an experimental set of FORCs;
- Parameters identification tools for all implemented models (including the user-defined models);
- Thermal relaxations and after-effect phenomena in both scalar and vector hysteretic systems;
- Noise transfer through hysteretic systems;
- Iterative simulations in which one or more parameters are changing according to the specifications of the user.

The rest of the article is organized as follows. In section 2 we introduce the main hysteresis models that come by default implemented in HysterSoft©. The next two sections focus on how to perform simulations with scalar and vector models of hysteresis, respectively. Section 5 presents how users can define new models of hysteresis in HysterSoft© (user-defined models), section 6 presents how to compute the scalar and vector Preisach distribution function from FORCs, section 7 presents how to use HysterSoft© as a library and how to work with files, and section 8 concludes.

2. Hysteresis models predefined in HysterSoft©

The main models of hysteresis implemented in HysterSoft© are presented in Table I. A detailed description of the constitutive equations for each model, parameter definition, and numerical implementation can be found on the references mentioned in the Table for each model. Users can also easily implement and employ additional models to the program by using predefined templates.

Table 1. Scalar and vector models implemented in HysterSoft©.

	Scalar models	Definition	Can be used in vector hysteresis simulations?	Can be used in inverse modeling?
1	Backslash operator	[17]	yes	yes
2	Bouc-Wen model	[8-10, 18-20]	yes	yes
3	Coleman-Hodgdon model	[14, 15, 21-25]	yes	yes
4	Elastic-plastic (stop) operator	[17]	yes	yes
5	Energetic model[a]	[16, 26-30]	yes	yes
6	Jiles-Atherton model	[11-13, 31-33]	yes	yes
7	Langevin model[b]	[34]	yes	yes
8	Limiting-loop proximity model	[35]	yes	yes
9	Preisach model[c]	[3-7, 36-38]	yes	yes
10	Rectangular loop operator	[17, 39]	no	no

[a] Also known as the Hauser model.

[b] Langevin model is not considered a model of hysteresis however it often represents the building block of other models of hysteresis such as the Jiles-Atherton model

[c] Preisach model can be defined either as a model in which the Preisach function is defined analytically or point-by-point, on a mesh.

3. Simulations using scalar models of hysteresis

HysterSoft© can perform different types of simulations by using the modules that are implemented in the program. These types of simulations can be performed for any hysteresis model, predefined or user-defined which justifies in part the popularity of the software. In this way, users need to concentrate only on developing the governing equations of the hysteresis model and not focus on adjusting it the model to a particular application. A module can be selected from the main combo-box in the scalar model window.

3.1. Modules available for scalar models of hysteresis

Module 'Input and output defined using the GUI'. Using this module, users can simulate hysteresis curves be varying the input in the case of direct hysteresis modeling or output in the case of inverse hysteresis modeling using the track bars from the graphical user interface. This module can be used to simulate only static hysteresis curves and visualize the current hysteretic state (e.g. the Preisach plane) in real time. Figure 1 presents such a simulation performed using the Preisach model.

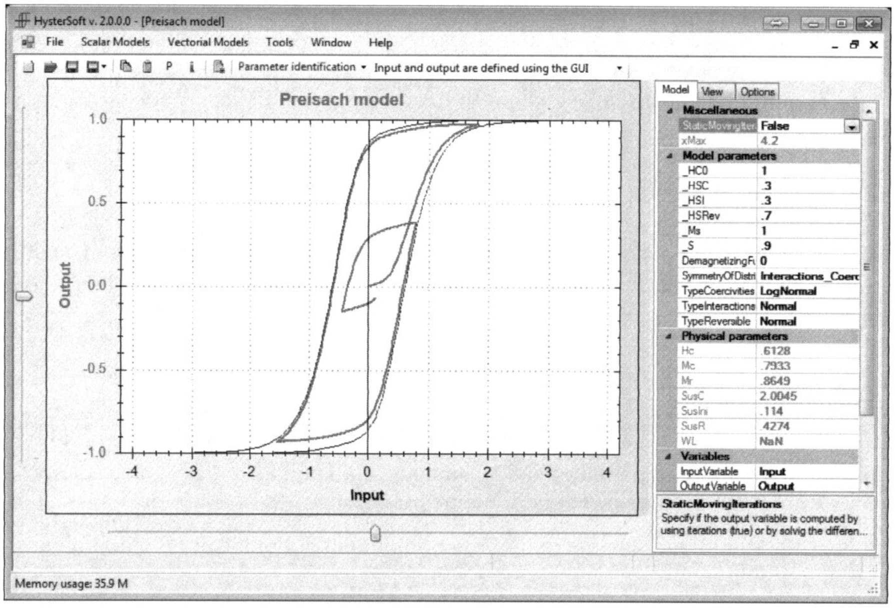

Figure 1. Tracing hysteresis curves using module 'Input and output defined using the GUI' in the case of the Preisach model. The Preisach distribution is normally distributed along the interaction axis and log-normally distributed along the coercivity axis. The reversible component is also assumed to be normally distributed along the $\alpha=\beta$ axis.

3.1.1 Module 'Input defined from file'. Using this module, users can simulate hysteresis curves using the values of the input (or output in the case of inverse modeling) defined in a file or in the text editor. The input (or output) should be defined one value per line. HysterSoft© will parse these values and compute the hysteretic curves. The user can use symbolic notations for the main physical parameters like in the following listing:

```
//sample input file
0
P1
-Hc
```

```
100
```

In the above example HysterSoft© will start from 0, apply an input equal to the value of user-defined parameter `P1` (which is set in the options property grid), then another value equal to minus the coercive field of the material, and finally 100. The final values of the input and output are stored in the `FinalX` and `FinalY` variables. This module can be used in loop simulations, but cannot be used to perform dynamic simulations.

3.1.2. Module 'Input defined analytically (dynamic hysteresis)'. Using this module, users can compute the output variable if the input is defined analytically. The input signal should be defined using the `InputSignal` editor in C#. When defining the input signal it is convenient to use variables such as `tMin`, `tMax`, `Hc`, `P1`, `P2`, `P3`, `P4`, etc. in order to easily modify the signal from the property grid. It is also convenient to use such variables when performing loop simulations. Figure 2 presents a rate-dependent simulation performed using the Jiles-Atherton model.

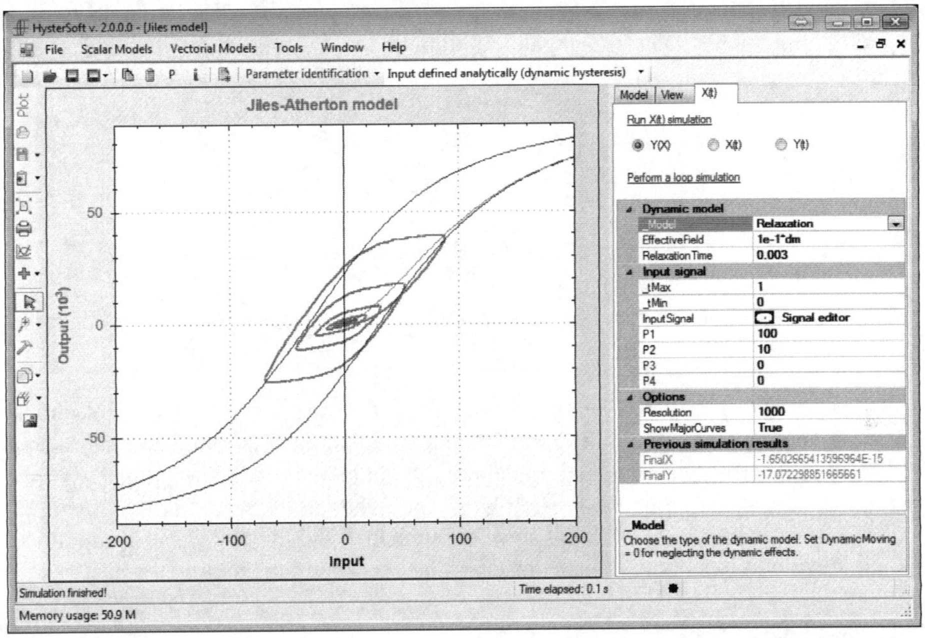

Figure 2. Sample rate-dependent simulations performed with the Jiles-Atherton model for a soft magnetic ferrite. The red line shows the rate-dependent hysteresis loop, while the blue line denotes the "static" major hysteresis loop. The model parameters are a = 41.8, α = 0, c = 0.15, k = 39.3 and $y_{sat} = 10^5$.

This module allows performing rate-dependent simulations by setting the `EffectiveField` or `RelaxationTime` variables. When the `EffectiveField` variable is set, HysterSoft© will use the effective field approximation, and compute output y using

$$y = \hat{\Gamma} x_{eff}, \tag{1}$$

$$x_{eff} = x + F(y, \dot{y}), \tag{2}$$

Journal of Physics: Conference Series **585** (2015) 012002 doi:10.1088/1742-6596/585/1/012002

where x is the input, x_{eff} is the effective value of the input, $\hat{\Gamma}$ is the scalar hysteresis operator, and F is a function of the output variable y and of its derivative with respect to time, $\dot{y}$. Equations (1) and (2) represent a system of nonlinear equations that is solved to compute y. In the framework of the Preisach model, this system is a system of integro-differential equations, while in the framework of the Jiles, energetic and Hodgdon models it becomes a system of differential equations. When the `RelaxationTime` variable is set, HysterSoft© will use the relaxation time approximation and compute y using:

$$\dot{y} = -\frac{y(t) - \hat{\Gamma}x(t)}{\tau}, \tag{3}$$

where τ is a relaxation time parameter and $\hat{\Gamma}$ is any regular static (not rate-dependent) hysteresis operator.

The relaxation time should be a real number, while the effective field should be a string variable that defines the effective field as a function of the normalized rate (i.e. time derivative) of the output variable. For instance, by setting the `EffectiveField` variable to

```
1.2*dm
```

HysterSoft© will use a total input equal to

$$x_{eff}(t) = x(t) + 1.2\frac{\dot{y}(t)}{y_{sat}} \tag{4}$$

Rate-dependent simulations are particularly convenient for simulating frequency-dependent hysteresis.

3.1.3. Module 'FORCs computations'. Using this module, users can compute a set of first-order reversal-curves. One can define the maximum applied input (`XMax`), the number of reversal curves (`Reversals`), and the number of points per curve (`Resolution`) in the property grid. Dynamic models cannot be used when computing FORCs. If one needs to compute FORCs using a dynamic model it is recommended to define the input signal analytically and use the previous module (0).

Figure 3 presents a sample FORCs computation using the limiting-proximity model.

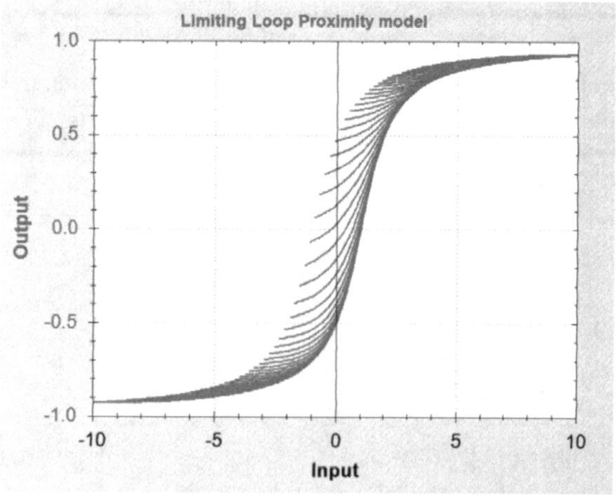

Figure 3. A set of 100 FORCs computed using the limiting loop proximity model. The model parameters are h_0 = 1, H_c = 1, y_{sat} = 1, and z = 0.9.

3.1.4. Module 'Thermal relaxation'. Using this module, users can perform simulations involving thermal relaxation. Thermal relaxation simulations are performed by adding a "thermal noise" to the input of a hysteresis model. The initial state can be obtained in two ways. The first way is to bring the hysteretic system in a given state by running a simulation with any other module (for instance any of the first 3 modules), then copying the hysteretic state, and using it as the initial state for the thermal relaxation simulation. The second way to set the initial state is to use the `FirstFieldToApply` and `SecondFieldToApply` variables as follows: before any thermal relaxation simulation HysterSoft© starts from the zero anhysteretic state and applies an input equal to `FirstFieldToApply`, then another input equal to `SecondFieldToApply`, and, then, performs the thermal relaxation simulations. If one does not want to use these variables one needs to set the `FirstFieldToApply` and `SecondFieldToApply` to `NaN`.

One can select different types of noise to use in the thermal relaxation including Ornstein-Uhlenbeck, Gaussian, Laplace, Cauchy, and uniformly distributed noise by setting the `TypeOfNoise` variable. For each type of noise one can set the magnitude of the noise and different other parameters characteristic to the noise selected. In addition, one can change the spectrum of the noise by changing the `NormalizeInputSpectrum` variable. By default the `NormalizeInputSpectrum` is set to 1, which means that the input noise is multiplied by 1 (i.e. left unchanged), however the user can change this variable to a frequency dependent function in order to modify the color of the noise. In this way, one can use colored noise inputs such as pink, blue, or violet noises as input noises.

One can also define the number of total averages after which the expected values of the output variable is computed, the experiment time, and the time step of the thermal noise. After the simulation is performed, HysterSoft© will set the `ViscosityCoefficient` variable to the value of the viscosity coefficient. This module can be used in loop simulations to analyze more complex phenomena such as data collapse, field-cooled or zero-field cooled processes.

Figure 4 presents a sample thermal relaxation simulations performed with the energetic and Preisach models. In both simulations we used a system with saturation equal to 1.62×106 and coercivity approximately equal to 8. The simulations started with the initial anhysteretic state after which the input was increased to 10 we waited for the thermal relaxation to occur.

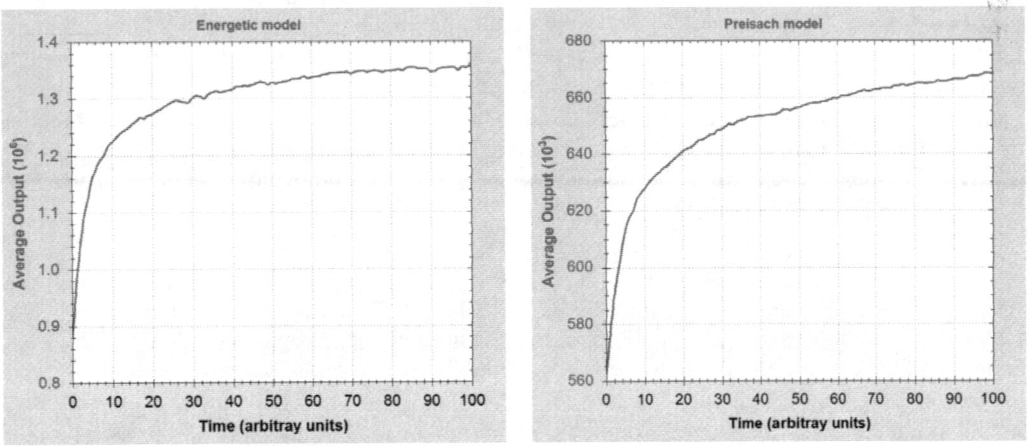

Figure 4. Simulations of thermal relaxation in the energetic model (left) and Preisach model (right), in which the input noise is Ornstein-Uhlenbeck with b = 1, $\sigma = 5$, and $x_0 = 0$. The parameters for the energetic model are $c_r = 1.32$, $g = 8.24$, $h = 0.217$, $k = 15.9$, $y_{sat} = 1.62\times10^6$, $N_e = 10^{-6}$, and $q = 15.3$., In the case of the Preisach model we used a normal distributions along interaction field and coercivity axes with $h_{c0} = 13$, $h_{\sigma c} = 6$, $y_{sat} = 1.6\times10^{6.}$ and $h_{\sigma i} = 8$, while the reversible distribution (along the $\alpha=\beta$ axis) was normal with $h_{\sigma rev} = 15$.

3.1.5. Module 'Noise passage analysis'. Using this module, users can analyze the noise passage characteristics through hysteretic systems [40-45]. This module allows one to compute the output spectrum of any model of hysteresis if the input spectrum is given. The input spectrum can be of any of the types described in the previous module. This module can be used in loop simulations.

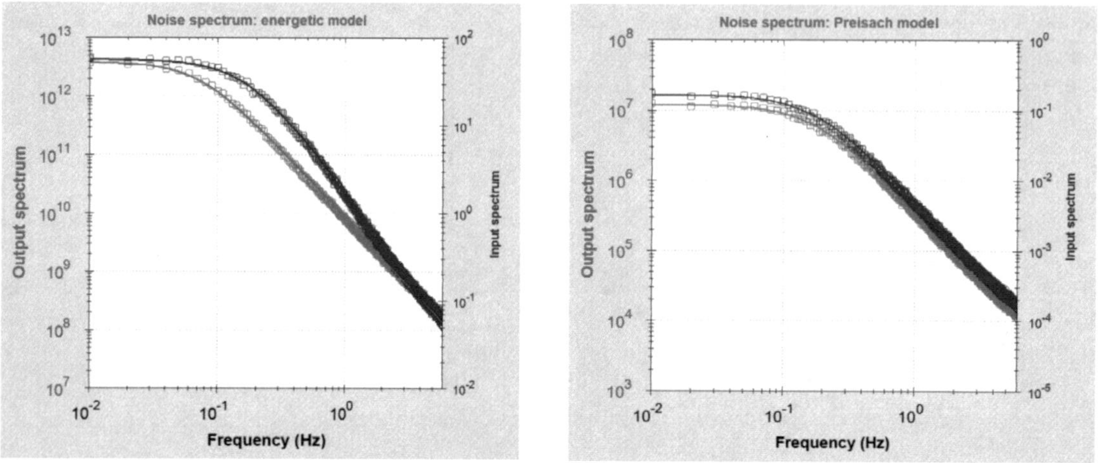

Figure 5. Noise passage analysis performed with the energetic model (left) and Preisach model (right), in which the input noise is Ornstein-Uhlenbeck with the same parameters as in the Figure 4. The model parameters are also identical to the ones used in Figure 4.

3.2. Loop simulations

Loop simulations refer to simulations that are performed multiple times, each time for a different set of model parameters or input signal. For instance, such simulations are convenient to perform when computing the anhysteretic curve, when computing hysteresis curves for different values of model parameters, or when simulating the thermal relaxation, noise passage, or stochastic resonance for different parameters of the input noise.

Loop simulation can be performed using the Loop Simulation Editor. By default, HysterSoft© performs two embraced iteration loops for two different parameters. The parameters and the values that these parameters should take during each simulation can be set in the "Major loop" and "Minor loop" tabs of the "Loop simulation" editor. The user can specify what variables to save on the hard disk, what curves from the output window to save, and what image files to produce after each iteration. The file names can be defined using parameters between two `$` characters. For instance

```
Output_Hc_$Hc$.dat
```

will be expended by replacing `$Hc$` with the current value of the coercive field.

3.3 Inverse modeling

Hysteresis models provide in general a procedure to compute the output variable $y(t)$ as a function of the input variable $x(t)$. In practical applications, it is often necessary to solve the inverse problem, in which one needs to compute the input variable that results in a given output. This is usually required when solving optimization problems or trying to design a hysteresis system.

Since hysteresis is a nonlinear and multi-valued function, the inverse problem might have no solution, one solution, or multiple solutions. One example is to invert the rectangular hysteresis operator, which, depending on the output variable, has no solution (if y is different from 1 or -1) or infinity of solutions (if y is either 1 or -1). However, quite often it is possible the invert a hysteresis model relatively easily using the technique described below.

Suppose we have a hysteresis model defined as

$$y(t) = \hat{\Gamma} x(t) \tag{5}$$

where $\hat{\Gamma}$ is the hysteresis operator. We introduce the differential susceptibility of the system as

$$\dot{y}(t) = \hat{\chi} \dot{x}(t) \tag{6}$$

where $\hat{\chi}$ is the susceptibility. We suppose that such susceptibility operator exists, is finite, and has an inverse operator $\hat{\chi}^{-1}$ such that $\hat{\chi}^{-1}\hat{\chi} = 1$. In this case:

$$\dot{x}(t) = \hat{\chi}^{-1} \dot{y}(t) \tag{7}$$

In the case of scalar hysteresis models, $\hat{\chi}$ is the differential susceptibility and $\hat{\chi}^{-1} = 1/\hat{\chi}$. If the hysteresis model is a differential model of hysteresis, the computational overhead when solving the direct problem (6) is equal to the computational overhead when solving the inverse problem (7). Most scalar models of hysteresis used in this book can be written in the form of (6). Hence, it is possible to invert these models by using (7).

HysterSoft© implements inverse modeling for all the hysteresis models for which the differential susceptibility is defined, including the Preisach and the energetic models. Inverse modeling in HysterSoft© can be performed only by using the first two modules ('Input and output defined using the GUI' and 'Input defined from file'.). If the values of the output are specified in an external file, the `ModelingType` variable should be set to `InverseModeling`.

In the case of user-defined models, inverse modeling can be performed only if the model implements the `Susceptibility` or the `ChangeY` functions (see section 5).

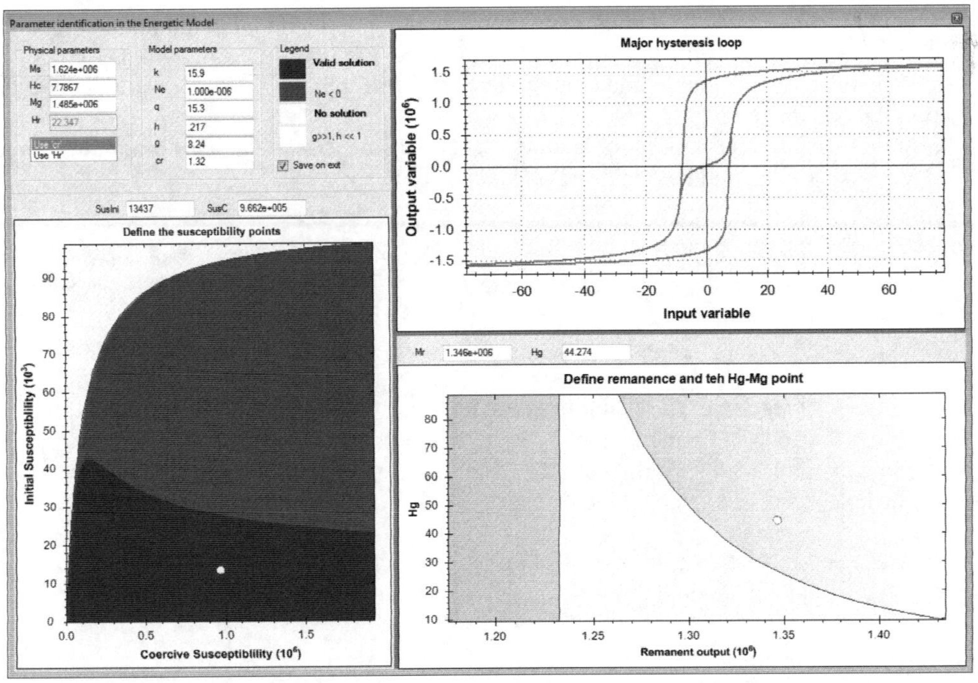

Figure 6. Non-iterative determination of parameters for the Energetic model [47].

3.4. Parameter identification tools

HysterSoft© implements three types of methods to determine the model parameters [33, 38, 46]:

1. evolutionary identification techniques such as swarm optimization, genetic algorithms, and the Nelder-Mead method;
2. the Levenberg–Marquardt algorithm;
3. iterative techniques.

In addition, HysterSoft© implements model specific, parameter identification tools. For instance Figure 6 shows the non-iterative parameter identification tool in the case of the energetic model) [47]. The values of the physical parameters that are be used in the identification problem (such as the coercive field, the output saturation, the remanence, initial susceptibility, etc.) can be specified on the graphical user interface.

In the case of user-defined models, the model parameters that appear in the parameter identification editor should be identified using the `[IdentificationModelParameter]` attribute.

4. Simulations using vector models of hysteresis

By default, HysterSoft© adds a two-dimensional vector model of hysteresis for most scalar models. The vector model is defined using superposition described in [39, 45, 48, 49]. For instance, two-dimensional vector models of hysteresis are constructed as a two-dimensional superposition of scalar models of hysteresis

$$\boldsymbol{y}(t)=\int_{-\pi/2}^{\pi/2}\hat{\varphi}\hat{\Gamma}_{\varphi}\left[\boldsymbol{x}(t)\cdot\hat{\varphi}\right]d\varphi, \tag{8}$$

where $\hat{\varphi}$ is the unit vector along the direction specified by polar angle φ, $\boldsymbol{x}$ is the vector input variable, and $\hat{\Gamma}_{\varphi}$ is the scalar hysteresis operator along direction φ. $\hat{\Gamma}_{\varphi}$ can be any scalar hysteresis operator such as the Preisach, Jiles-Atherton, energetic, or other scalar models. It should be mentioned that model (8) was initially introduced in [4] for the Preisach model. Due to the high computational cost required to integrate (8), this equation is relatively difficult to implement in real-time simulations for the Preisach model, however, it becomes more manageable when used with other models such as the energetic of Jiles-Atherton models.

Similar to the case of scalar models of hysteresis, HysterSoft© has a number of modules that can be used to perform vector simulations. These modules are summarized below.

4.1. Module 'Input and output defined using the GUI'. Simulate vector hysteresis curves by varying the input using the buttons on the graphical user interface (these buttons can also be pressed using the arrow buttons on the keyboard).

4.2. Module 'Input defined from file'. Simulate hysteresis curves by using the input from a file or from a textbox. Both the x and y components of the input should be defined one the same line. HysterSoft© will parse these values and compute the output variable. Notations for various parameters can also be used like in the next example

```
//sample input file for vector simulations
0, 0
HcX, 0
HcX, HcY
P1, P2
0, -P3
```

Journal of Physics: Conference Series **585** (2015) 012002 doi:10.1088/1742-6596/585/1/012002

In this example HysterSoft© starts from point (0, 0), than applies (`HcX, 0`), (`HcX, HcY`), (`P1, P2`), and finally (`0, -P3`). The values of `HcX`, `HxY`, `P1`, `P2`, `P3`, etc. will be expended to the numerical values defined by the user in the property grid. The final value of the input and output components after the simulation are given by the `FinalX_x`, `FinalX_y`, `FinalY_x`, and `FinalY_x` variables. These variables can be used in loop simulations.

This module can also be used in inverse modeling by setting the `ModelingType` variable to `InverseModeling`. In this case, the input file should specify the values of the output of the hysteretic system, and HysterSoft© will compute the corresponding input values.

Figure 7 presents sample two-dimensional simulations using the Preisach and Bouc-Wen models. In both simulations the input was increased along the x-axis from 0 to 100 (in the case of the Preisach model) and to 20 (in the case of Bouc-Wen model) and then rotated in the x-y plane. The parameters of both models are distributed uniformly on the unit circle to simulate an isotropic material.

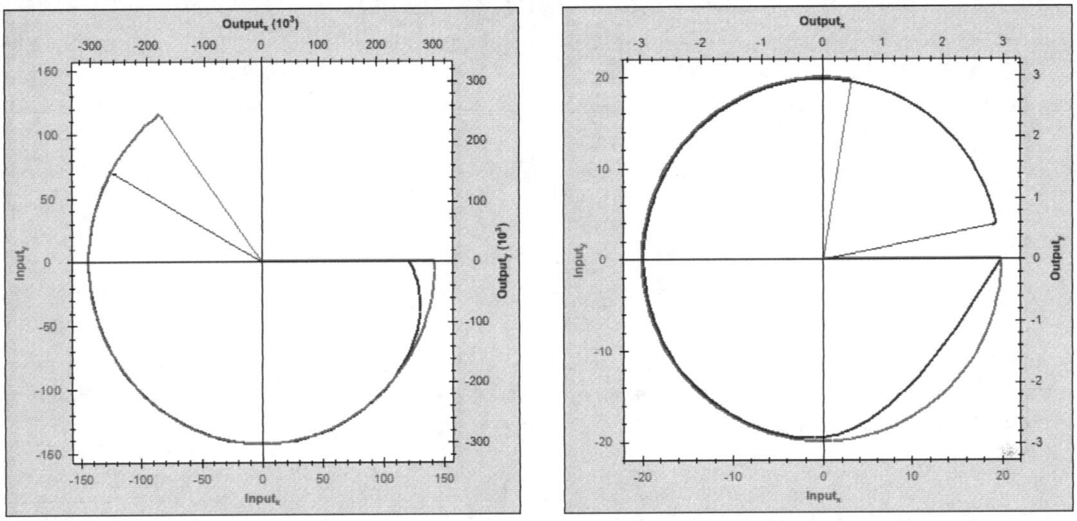

Figure 7. Two-dimensional vector simulations performed with the Preisach model (left) and Bouc-Wen model (right). In the case of the Preisach model we used a normal distributions along interaction field and coercivity axes with $h_{c0} = 100$, $h_{\sigma c} = 30$, $y_{sat} = 3.5\times10^{5\cdot}$ and $h_{\sigma i} = 30$, while the reversible distribution (along the α=β axis) was normal with $h_{\sigma rev} = 70$. The parameters of the Bouc-Wen model are $A = 2$, $\alpha = 0$, $\beta = 0.5$, $D = 1$, $\gamma = 0.1$, $k = 1$, and $n = 1.1$. The trajectory of the output variable is represented with blue line, while the trajectory of the output variable is represented with red line. The final arrow represents the position of the two-dimensional input and output at the last simulations point

4.3. Module 'FORCs computations'. Compute vectorial FORCs, by specifying the maximum applied input (`XMax`), the number of reversal curves (`Reversals`), the number of points per curve (`Resolution`), and the angle under which the FORCs are computed (`Angle`). These values can be all be specified in the property grid. Dynamic models cannot be used when computing vectorial FORCs.

4.4. Module 'Thermal relaxation'. Perform two-dimensional thermal relaxation simulations. The initial hysteretic state can be specified by using the copy and paste features. The user can also specify the magnitude of the noise, the number of averages, and the number of points in which the thermal relaxation is computed.

Journal of Physics: Conference Series **585** (2015) 012002 doi:10.1088/1742-6596/585/1/012002

Figure 8 presents the results of a two-dimensional simulation of the thermal relaxation using the energetic model. The left figure shows how the initial state was obtained using module 'Input and output defined using the GUI': starting from the anhysteretic point we increased the input along the x-direction to 5×10^5 and rotated it by 60 degrees in counterclockwise direction. The figure on the right shows the trajectory of the output which started in the initial state 1 and ended in final state 2 after 100 iterations.. The input noise was assumed normally distributed standard deviation of 1×10^4. The trajectory of the output is represented with a blue line, while the input noise with represented in red.

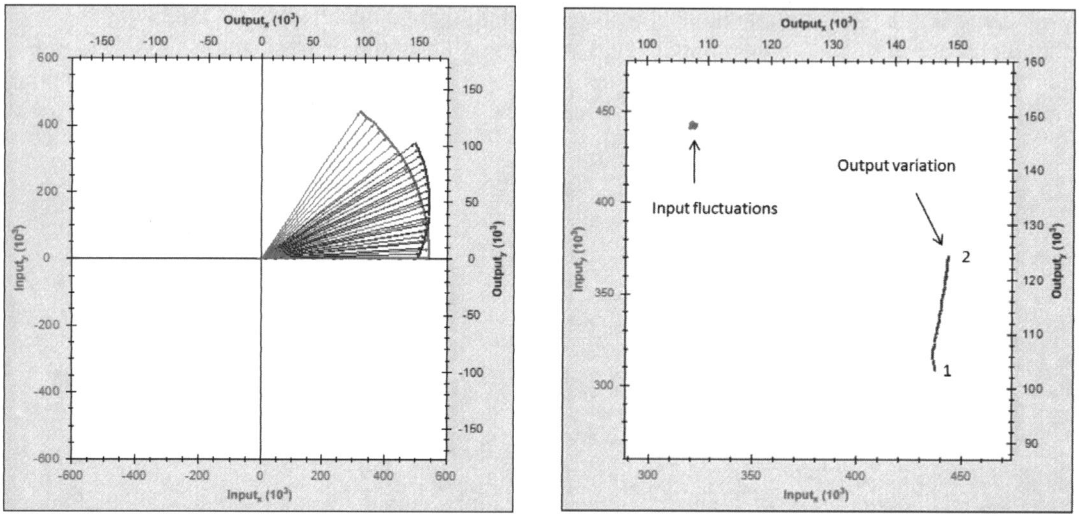

Figure 8. Two-dimensional thermal relaxation in the energetic model. The left figure shows how the initial state of the system is obtained. The right figure shows the effect of a normally distributed input noise (represented with red) on the output vector (represented with blue).

5. Defining new hysteresis models

New scalar and vector hysteresis models can be easily defined in HysterSoft©. To define a new scalar model, start the "New Scalar Model" editor from the "Scalar Models" menu. Using this editor one can define the equations of the new model, properties that should appear in the property grid, default values for the model parameters, the parameters should appear in the parameter identification tool, etc. The easiest way to learn how to define new hysteresis models is to use examples that come with the installation kit.

The code of the new model should be written in C# and should contain the definition of the model as a class in C#. This class should be inherited from the `UserDefinedModel` class (which is internally implemented in HysterSoft©). Any user-defined model should define at least one of the following two functions

```
double Susceptibility(double x, double y, int sensOfVariation)
```

or

```
double ChangeX(double x)
```

The first function defines the susceptibility of the model, while the second describes how the output variable changes when the input becomes equal to `x`). The first function is convenient for use in differential models of hysteresis, while the second one in algebraic models. Once any of the above

Journal of Physics: Conference Series **585** (2015) 012002 doi:10.1088/1742-6596/585/1/012002

functions is implemented, the user can perform any types of simulations presented above, including thermal relaxation and noise passage analysis.

Now let us look at an example that comes by default in HysterSoft©. This example implements the Jiles-Atherton model.

```
public class MyModel : UserDefinedModel
{
    //a default constructor is optional
    public MyModel()
    {
        xMax = 100;
    }

    //Define the differential susceptibility
    public override double Susceptibility(double x, double y, int sensOfVariation)
    {
        double l, dl;
        l = SpecialFunctions.Langevin(x+alpha*y)-y;
        dl = SpecialFunctions.dLangevin(x+alpha*y);

        if (sensOfVariation*l <= 0)
            return c*dl;

        //return the differential susceptibility
        return (1-c)*l/(k*(1-c)*sensOfVariation-alpha*dl)+c*dl;
    }

    //Add some properties on the property grid
    private double a = 20;
    [IdentificationModelParameter(1, 100)]
    public double _a { get{return a;} set{a=value;} }

    private double c = 0.1;
    [IdentificationModelParameter(0, 1)]
    public double _c { get{return c;} set{c=value;} }

    private double k = 40;
    [IdentificationModelParameter(1, 100)]
    public double _k { get{return k;} set{k=value;} }

    private double alpha = 0;
    [IdentificationModelParameter(0, 1e-5)]
    public double _alpha { get{return alpha;} set{alpha=value;} }
}
```

The `Susceptibility` function is mandatory (because we did not implement the `ChangeX` function). The rest of all other functions and properties are optional, including the default constructor.

The `[IdentificationModelParameter(...)]` attribute for properties is also optional and tells HysterSoft© that the given property defines a model parameter that should appear in the Parameter Identification Tool. The two parameters of the `IdentificationModelParameter` attribute

Journal of Physics: Conference Series **585** (2015) 012002 doi:10.1088/1742-6596/585/1/012002

denote the minimum and maximum values within which HysterSoft© can search during the parameter identification.

Functions `SpecialFunctions.Langevin` and `SpecialFunctions.dLangevin` are internally defined in HysterSoft© and they compute the Langevin function and the derivative of the Langevin function. The user can use his or her own implementation for these functions.

The `ToString` function defines a name for the current model.

Next let us look at another example that implements the Langevin model, which is an algebraic model.

```
public class MyModel : UserDefinedModel
{
    public override double ChangeX(double x)
    {
        return Ms*SpecialFunctions.Langevin(x/3);
    }
}
```

Notice that HysterSoft© will always define a variable `Ms`, which is by default equal to 1. Since hysteresis is a history dependent phenomenon, HysterSoft© allows users to recall the previous values of the input and output variables. The following predefined variables can be used from anywhere inside the class definition, including from inside functions `Susceptibility` and `ChangeX`:

- `state.X` – the last value of the input
- `state.Y` – the last value of the output
- `state.XReversal` – the last value of the input reversal
- `state.YReversal` – the last value of the input reversal
- `state.pastX[i]` – the previous `i`-th value of the input
- `state.pastY[i]` – the previous `i`-th value of the output

Notice in the last example that HysterSoft© will not be able to perform inverse modeling simulations because the class does not implement the `Susceptibility` or the `ChangeY` functions.

6. Computation of the Preisach distributions using first-order reversal-curves (FORCs)

FORCs have become an essential tool to characterize hysteretic systems [36, 49-52]. In magnetic materials they provide information about the magnetic interactions and particle coercivities and are often used as the "fingerprints" of a given system.

6.1. Scalar FORCs Analysis

The "Scalar FORCs Analysis" tool in HysterSoft© can be used to compute Everett distribution, the reversible $R(\alpha)$ and irreversible $P(\alpha,\beta)$ Preisach distributions if a set of first-order reversal-curves (FORCs) is provided. The set of FORCs should be given as a 3-column file with the values of the reversal field, current field, and current output for each curve, as in the example below:

```
10 10 1       : this is the first curve which consists of one point
9 9 0.99      : here starts the second reversal-curve which consists of 2 points
9 10 1
8 8 0.98      : here starts the third reversal-curve which consists of 3 points
8 9 0.99
8 10 1
...
```

```
-10 -10 -1    : here starts the last reversal-curve
-10 -9 -1
...
-10 10 1
```

The user can specify any number of reversal curves, each having any number of points. The reversal fields and the points on each reversal curve can be distributed non-uniformly, however, in order to increase the accuracy of computations it is strongly recommended to use uniform distributions. HysterSoft© will automatically detect the number of reversal curves and points on each curve. FORCs should be given as ASCII files with extension `.forcs`.

The Everett distribution can be saved as a `.everett` file, while the computed Preisach distribution as a `.preisach` file. The structure of the `.preisach` file is similar to the structure of the `.forcs` file and consists of 3 columns giving the values of the reversible component of the Preisach distribution $R(\alpha)$ when $\alpha = \beta$ and the irreversible component of the Preisach distribution irreversible $P(\alpha,\beta)$ when $\alpha > \beta$. This is a sample `.preisach` file

`10 10 0.01`	: $R(10) = 0.01$
`9 9 0.02`	: $R(9) = 0.02$
`9 10 0`	: the first point of the irreversible component $P(9,10) = 0$
`8 8 0.03`	: $R(8) = 0.03$
`8 9 0.1`	: the next point of the irreversible component $P(8,9) = 0.1$
`8 10 0.02`	: the next point of the irreversible component $P(8,10) = 0.02$
`...`	
`-10 -10 0.1`	: $R(-10) = 0.1$
`-10 -9 -0.1`	: $P(-10,-9) = 0.1$
`...`	
`-10 10 0.1`	: $P(-10,10) = 0.1$

The `.everett` and `.forcs` files can be loaded by the Preisach model with discrete Preisach distributions and used to perform frequency-dependent simulations or thermal and noise analysis.

6.2. Vector FORCs Analysis

To identify the Preisach distributions of the vector Preisach model, HysterSoft© allows users to load multiple `.forcs` files and vector distributions. This can be done using the "Load FORCs Analysis" and "Load Multiple FORCs" tools. The FORCs in each angular direction need to be specified in different files and loaded in the software. HysterSoft© can compute the Preisach distribution function for both two-dimensional and three-dimensional FORCs in the case of isotropic hysteretic systems, however it can compute only the Preisach diagram for only two-dimensional FORCs in the case of anisotropic systems. Notice that two-dimensional FORCs should be specified along each polar angle φ, while three-dimensional FORCs should be specified along each polar and azimuthal angles φ and θ.

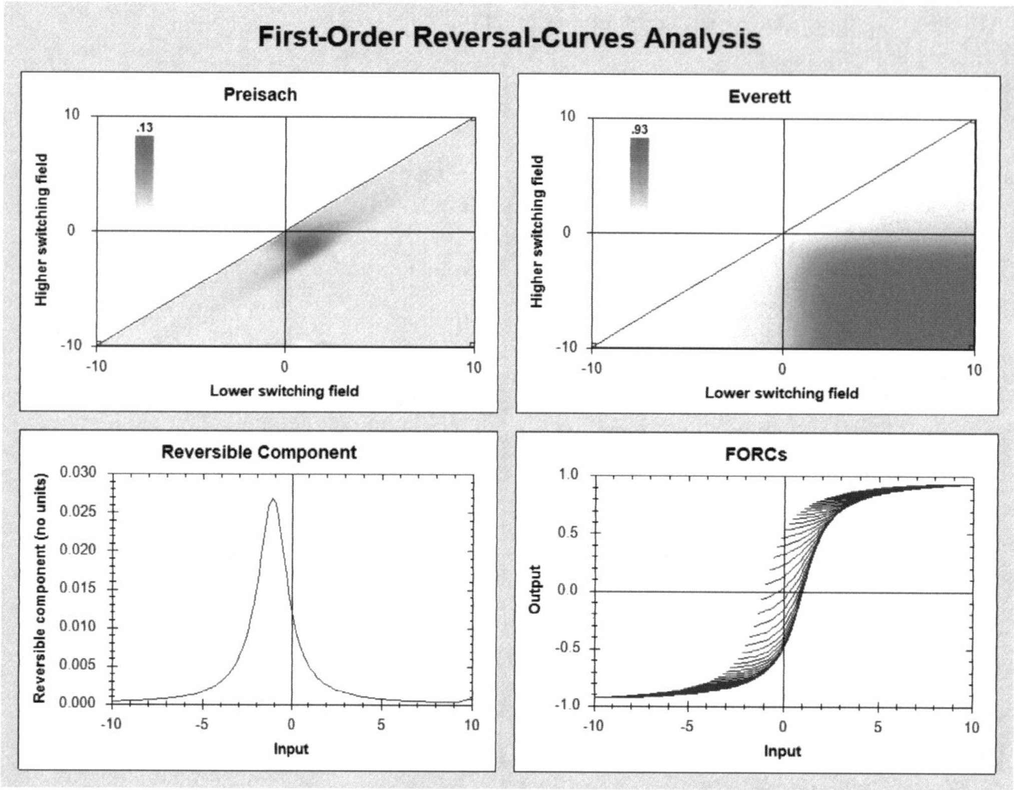

Figure 9. Computation of the reversible and irreversible components of the scalar Preisach distribution function from a set of FORCs.

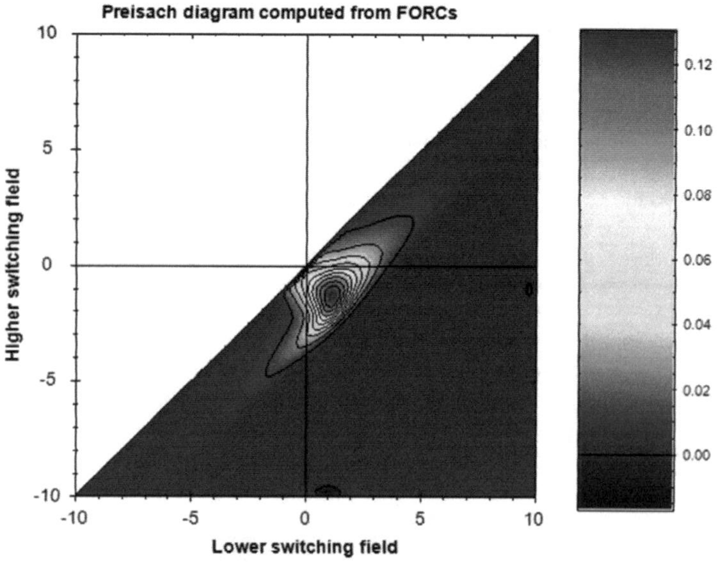

Figure 10. Representation of the Preisach distribution function computed in Figure 9.

Journal of Physics: Conference Series **585** (2015) 012002 doi:10.1088/1742-6596/585/1/012002

7. Other features

7.1. Using HysterSoft© as a library

Most HysterSoft© functions can be called from other Windows applications (such as Matlab, C, Fortran, Java, etc.). There are two ways to call HysterSoft© functions from other programs:

- load the HysteresisLibrary.dll file as a .NET assembly, create a desired model, and call the public functions and properties (recommended)
- use HysterSoft© as a COM object (not recommended; this way is obsolete and is kept for compatibility with old versions of HysterSoft©).

Examples of how to call HysterSoft© from Matlab can be found in the /Matlab subdirectory of the installation directory of HysterSoft©.

7.2. Saving the model parameters

The parameters of any model of hysteresis can be saved in .xml files that later can be re-loaded and used in other simulations. By default, HysterSoft© assigns the `.hyst` file extension to a model parameter file. Models parameters files are in general compatible from one version of HysterSoft© to another.

7.3. Perform temperature and stress dependent simulations

Temperature and mechanical stress change the shape of the hysteresis loops in most magnetic and electric hysteretic systems. For instance, in the case of magnetic systems, increasing the temperature will usually decrease the value of the coercive field, output remanence and saturation, while increasing the stress will increase the output remanence and make the shape of the major loop more rectangular. For instance, Figure 11 presents the results of increasing stress on the major hysteresis loop computed using the energetic model.

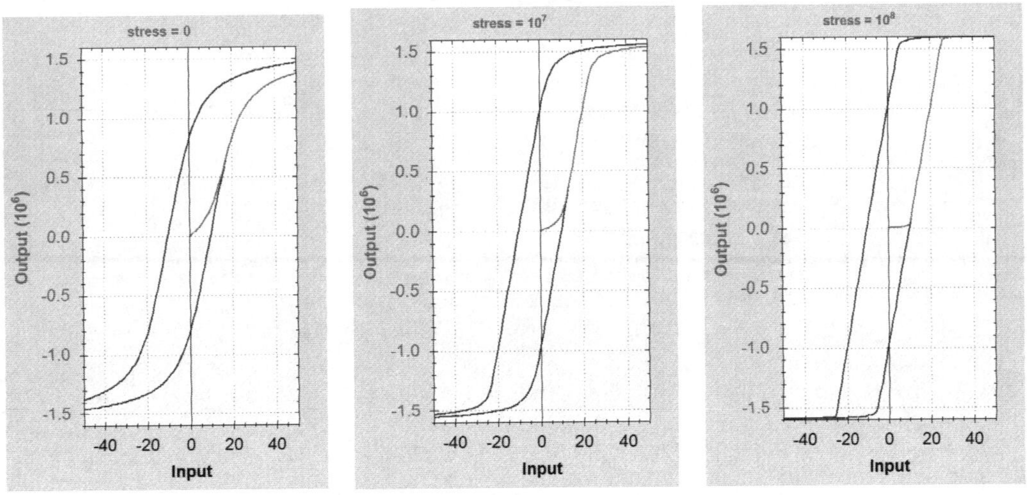

Figure 11. Major hysteresis loops computed using the energetic model at different values of the stress.

8. Conclusion

HysterSoft© is a computer software for the simulation of hysteresis and related phenomena in hysteretic systems. It provides a user-friendly simulation framework, in which various mathematical models of hysteresis can be implemented easily. A number of hysteresis models such as the backlash operator, the Bouc-Wen model, the Coleman-Hodgdon model, the stop (elastic-plastic) operator, the energetic model, the Jiles-Atherton model, the Langevin model, the limiting-loop proximity model, the Preisach model (including the cases in which the Preisach distribution is specified analytically or using discrete values defined on a mesh), and the rectangular-loop model are already implemented in the program. However, users can also implement and employ additional models by using predefined templates.

HysterSoft© can also be used to compute first-order reversal-curves (FORCs) diagrams, identify the model parameters from experimental data, conduct temperature and stress dependent simulations, perform noise passage analysis in hysteretic systems, etc. The program can be used as a library and called from other programs such as Matlab, Java, or C++. Finally, HysterSoft© can be used to define hysteresis models for the electric permittivity and magnetic permeability. These models can be imported directly in RandFlux© and used to simulate electronic circuits containing hysteretic inductors, transformers, and capacitors. A list of the most important features in HysterSoft© was also presented. Since the submission of this article the authors have finalized and published book [53], which also contains numerous simulation results based on HysterSoft©, as well as a short account of this simulation environment in the Appendix.

9. Acknowledgement

This paper was partially supported by the project "Progress and development through postdoctoral research and innovation in engineering and applied sciences - PRiDE - Contract no. POSDRU/89/1.5/S/57083", project co-funded from European Social Fund through Sectorial Operational Program Human Resources 2007-2013.

10. References

[1] HysterSoft, (Florida A&M University and Florida State University, Tallahassee, FL, Florida A&M University and Florida State University).
[2] RandFlux, (Florida A&M University and Florida State University, Tallahassee, FL, Florida A&M University and Florida State University), pp. User manual v. 0.6.
[3] Preisach F 1935 Z Phys **94** 277-302.
[4] Mayergoyz I 1991 *Mathematical Models of Hysteresis*. (Springer-Verlag).
[5] Visintin A 1995 *Differential Models of Hysteresis (Applied Mathematical Sciences)*. (Springer).
[6] Brokate M and Sprekels J 1996 *Hysteresis and Phase Transitions*. (Springer).
[7] Bertotti G and Mayergoyz I 2005 *The Science of Hysteresis*. (Elsevier).
[8] Bouc R 1971 Acustica **24** (1) 16-25.
[9] Wen Y K 1976 J Eng Mech Div-Asce **102** (2) 249-263.
[10] Duhem P 1897 Z. Phys. Chem. **22** 543.
[11] Jiles D C and Atherton D L 1983 IEEE Transactions on Magnetics **19** (5) 2183-2185.
[12] Jiles D C and Atherton D L 1984 Journal of Applied Physics **55** (6) 2115-2120.
[13] SPICE, (http://bwrc.eecs.berkeley.edu/classes/icbook/spice/, University of California at Berkeley).
[14] Coleman B D and Hodgdon M L 1987 Archive for Rational Mechanics and Analysis **99** (4) 375-396.
[15] Coleman B D and Hodgdon M L 1986 International Journal of Engineering Science **24** (6) 897-919.
[16] Hauser H 1994 Journal of Applied Physics **75** (5) 2584-2596.
[17] Visintin A 2005 *Mathematical models of hysteresis*. (Elsevier Inc.).
[18] Bouc R, presented at the Proceedings of the Fourth Conference on Nonlinear Oscillation, Prague, Czechoslovakia, 1967 (unpublished).
[19] Wen Y K 1976 J Struct Div-Asce **102** (5) 1160-1162.
[20] Ikhouane F and Rodellar J 2007 *Systems with hysteresis*. (John Wiley & Sons, Ltd.).
[21] Hodgdon M L 1988 IEEE Transactions on Magnetics **24** (6) 3120-3122.
[22] Hodgdon M L 1988 IEEE Transactions on Magnetics **24** (1) 218-221.

[23] Stancu A, Caltun O and Andrei P 1997 Journal De Physique IV **7** (C1) 209-210.
[24] Andrei P, Stancu A and Adedoyin A 2006 Journal of Optoelectronics and Advanced Materials **8** (3) 988-990.
[25] Andrei P and Adedoyin A 2010 Mathematics and Computers in Simulation **80** (6) 1045-1055.
[26] Hauser H 1994 Journal of Magnetism and Magnetic Materials **133** (1-3) 32-35.
[27] Hauser H 2004 Journal of Applied Physics **96** (5) 2753-2767.
[28] Andrei P, Stoleriu L and Hauser H 2000 Journal of Applied Physics **87** (9) 6555-6557.
[29] Grossinger R, Hauser H, Stoleriu L and Andrei P 2000 IEEE Transactions on Magnetics **36** (5) 3357-3359.
[30] Andrei P, Stancu A, Hauser H and Fulmek P 2007 Journal of Optoelectronics and Advanced Materials **9** (4) 1137-1139.
[31] Jiles D C and Atherton D L 1986 Journal of Magnetism and Magnetic Materials **61** (1-2) 48-60.
[32] Andrei P and Stancu A 1999 Journal of Magnetism and Magnetic Materials **206** (3) 160-164.
[33] Andrei P, Oniciuc L, Stancu A and Stoleriu L 2007 Journal of Magnetism and Magnetic Materials **316** (2) E330-E333.
[34] Kittel C 2005 *Introduction to Solid State Physics* 8th ed. (John Wiley & Sons, Inc.).
[35] de Almeida L A L, Deep G S, Lima A M N and Neff H 2003 IEEE Transactions on Magnetics **39** (1) 523-528.
[36] Andrei P, Caltun O and Stancu A 2006 Physica B-Condensed Matter **372** (1-2) 265-268.
[37] Ionita V, Petrescu L, Bordianu A, Tabara O, 2013 Adv. Electr. Comp. Eng. **13** (2), 121-126..
[38] Andrei P and Stancu A 2000 IEEE Transactions on Magnetics **36** (4) 1982-1989.
[39] Mayergoyz I 2003 *Mathematical Models of Hysteresis and Their Applications: Electromagnetism*. (Academic Press).
[40] Adedoyin A, Dimian M and Andrei P 2009 IEEE Transactions on Magnetics **45** (10) 3934-3937.
[41] Dimian M, Adedoyin A, Gindulescu A and Andrei P 2009 IEEE Transactions on Magnetics **45** (11) 5231-5234.
[42] Dimian M, Adedoyin A, Gindulescu A and Andrei P 2010 IEEE Transactions on Magnetics **46** (2) 266-269.
[43] Dimian M and Andrei P 2011 Journal of Applied Physics **109** (7) 07D330-333
[44] Dimian M, Andrei P, Manu O and Popa V 2011 IEEE Transactions on Magnetics **47** (10) 3825-3828.
[45] Andrei P and Adedoyin A 2008 Journal of Applied Physics **103** (7) 07D913.
[46] Jiles D C, Thoelke J B and Devine M K 1992 IEEE Transactions on Magnetics **28** (1) 27-35.
[47] Andrei P and Adedoyin A 2009 Journal of Applied Physics **105** (7) 07D523.
[48] Stancu A, Stoleriu L and Andrei P 2007 Journal of Magnetism and Magnetic Materials **316** (2) E309-E312.
[49] Stoleriu L, Andrei P and Stancu A 2008 Journal of Applied Physics **103** (7) 07D923-923
[50] Stancu A and Andrei P 2006 Physica B-Condensed Matter **372** (1-2) 72-75.
[51] Stancu A, Andrei P and Stoleriu L 2006 Journal of Applied Physics **99** (8) 08D702.
[52] Stoleriu L, Postolache P, Andrei P and Stancu A 2010 Optoelectron Adv Mat **4** (4) 505-508.
[53] Dimian M, Andrei P 2014 *Noise-Driven Phenomena in Hysteretic Systems*. (Springer).

Numerical testing of homogenization formulas efficiency for magnetic composite materials

Adelina Bordianu, Lucian Petrescu, Valentin Ionita
Faculty of Electrical Engineering, University Politehnica of Bucharest, 060042, Romania

E-mail: adelina.bordianu@upb.ro

Abstract. Magnetic composite materials are used for multiple applications. The macroscopic behavior of the material is influenced by its microscopic properties. Sometimes it is difficult to model the complex structure of the composite because the properties differ at the microscopic scale. Some simplifying assumptions are often needed, one of them being the homogenization of the material. In this paper two different composite materials were analyzed using a finite element software (COMSOL Multiphysics©). Both materials have a non-magnetic matrix with inclusions made of paramagnetic material for one of them, and ferromagnetic material for the other. The 3D numerical simulations were made for different particle concentrations and for different applied fields. The homogenization was implemented using two different formulas: Maxwell Garnett and Bruggeman. Numerical comparison was made between the magnetic properties of non-homogeneous materials and homogeneous ones showing each formula's efficiency for different cases.

1. Introduction

Starting from the middle of the 20th century the composite materials experienced an enormous development in multiple areas of the industry. Magnetic composite materials can be used in various applications, such as electromagnetic shielding, electrical machines, biomedical applications (for example, magnetic resonance imaging contrast enhancement or magnetic drug delivery), magnetic sensors [1] ÷ [5] and many other branches of the industry.

In the past years more and more scientists and research companies started developing new composite materials capable of improving the properties of classical materials. Magnetic composite materials can be described as a combination of minimum two materials: the matrix and the inclusions. The large number of combinations between the matrix and inclusions, imposes an important study of the structure and properties of the created composite. The continuous research and development of the composite materials leads to complex structures that cannot be analyzed easily. The solution in these cases is to make some simplifying assumptions. For example, all the material inclusions are considered being exactly the same from the geometrically point of view and perfectly embedded in the matrix. Other assumption, that was used successfully, is the homogenization technique. The idea behind this technique is to replace the complex structure of the real material with a homogeneous one that is able to describe correctly all the material properties (e.g. material conductibility, permeability, permittivity). The real material properties differ at the microscopic scale and it is important to know all these properties in order to describe correctly the material. By using the homogenization technique the medium becomes homogeneous, the properties are the same in each point and the macroscopic rules and formulas can be applied with success. One of the immediate results of using the homogenization is that the time needed to obtain the desire solutions is reduced.

Multiple homogenization techniques exist in technical literature. Some of them are based on analytical formulas (such as Maxwell Garnett, Bruggeman, Clausius-Mossotti), others based on different computational techniques (e.g. finite element method or finite difference method) [6] ÷ [14].

The object of this paper is to analyze, from the numerical point of view, the efficiency of Maxwell Garnett and Bruggeman homogeneisation formulas, implemeted in a three dimensional software, for two different materials.

Maxwell Garnett homogenization rule [15] is one of the first techniques used to describe the permeability of composite materials. It can be successfully used in describing materials with spherical and ellipsoidal inclusions but it is not limited to these applications alone. The initial formula was developed for describing the material permittivity, but can be used with good results for describing the permeability of the composite material. The Maxwell Garnett homogenization formula is [16]:

$$\frac{\mu_o - \mu_m}{\mu_o + 2\mu_m} = f \cdot \frac{\mu_i - \mu_m}{\mu_i + 2\mu_m} \tag{1}$$

where indexes *o, m* and *i* indicate the homogeneous, matrix and inclusion magnetic permeability, and f is the volume fraction of the inclusions in the matrix.

Bruggeman extended the previous model by taking into consideration the interaction between particles. The Bruggeman homogenization formula is [15]:

$$(1-f)\cdot\frac{\mu_m - \mu_o}{\mu_m + 2\mu_o} + f \cdot \frac{\mu_i - \mu_o}{\mu_i + 2\mu_o} = 0 \tag{2}$$

where indexes *o, m* and *i* indicate the homogeneous, matrix and inclusion magnetic permeability, and f is the ratio between the inclusion and the matrix volume.

By performing the numerical comparison between the magnetic properties of the real composite materials and homogeneous ones each formula's efficiency can be pointed for different concentrations (f values).

2. Investigated Materials

Two different composite materials were considered. The first one, Material A, is a composite material developed for biomedical applications, which has a non-magnetic matrix (cellulose) with inclusions made of paramagnetic material. The second one, Material B, is a soft magnetic composite material, also with a non-magnetic matrix, but with inclusions made of ferromagnetic material.

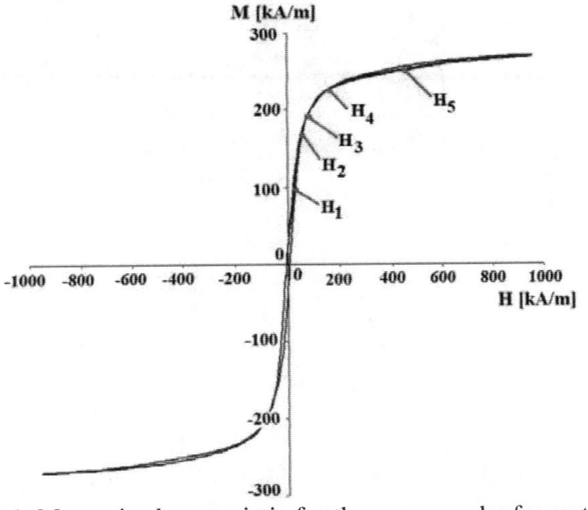

Figure 1. Magnetic characteristic for the pure powder for material A

The magnetic characteristics of the two materials samples were measured using the Vibrating Sample Magnetometer (VSM) device – the sample been a thin disk with 5 mm diameter. The characteristics for the spherical inclusions (the mean particle radius is approximately 20 nm) were obtained from the VSM results (figure 1) and they were used in the numerical approach in order to create a model which is very close to the real material.

3. Numerical approach

Using the finite element software COMSOL Multiphysics © [16], 3D models were made for both materials for the static case. A cube was chosen to model the composite material (figure 2). The inclusions were considered spherical in shape (inclusion radius = 20 nm) and were distributed randomly.

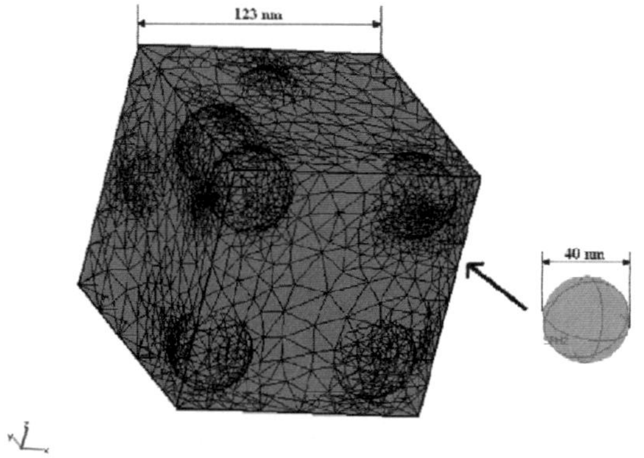

Figure 2. The 3D model used for the real material

Five different particle concentrations were considered: 50%, 40%, 30%, 20% and 9%. For each concentration, the number of inclusions was calculated – respectively 27, 22, 17, 11 and 5. Using the boundary conditions, a uniform magnetic field was applied in the x direction (figure 3). A refined mesh was chosen with approximately 100.000 finite elements for the composite material with the highest number of inclusions

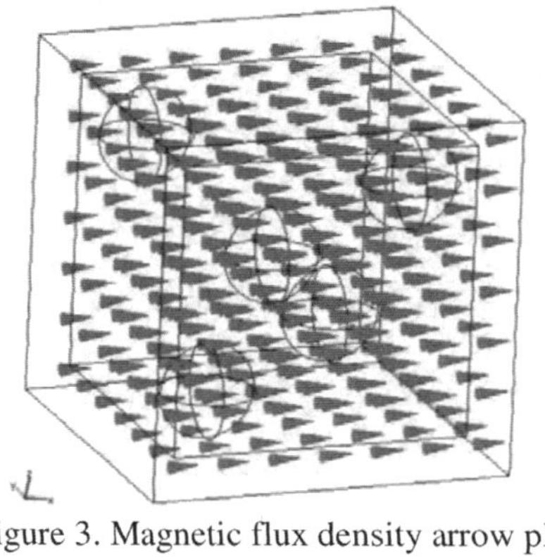

Figure 3. Magnetic flux density arrow plot

The matrix magnetic permeability is μ_m=1 for both materials. The permeability of the inclusions was chosen from the $\boldsymbol{B}=f(\boldsymbol{H})$ curve of the real material. Thus, for material A, five relative permeability values were chosen, each one corresponding to one different point from the *B-H* characteristic: μ_{r_P1} = 4.22, μ_{r_P2} = 3.34, μ_{r_P3} = 2.24, μ_{r_P4} = 1.44, μ_{r_P5} = 1.05. For material B, three magnetic permeability values were considered: μ_{r_P1} = 6225.13, μ_{r_P2} = 589.65, μ_{r_P3} = 99.76. Each permeability value imposes different values for the boundary conditions, which means different values for the applied magnetic field.

Numerical simulations were made for different particle concentrations: 9%, 30% and 50% for material A and 50%, 40%, 30%, 20% and 9% for material B. For each concentration three simulations were performed: the composite material and two homogeneous materials (where the material has the same relative permeability, in one case computed with Maxwell Garnett formula, and in the other case using Bruggeman). The results obtained were compared and different conclusions were made.

4. Results and discussions

The first step in the analysis is to compute the homogeneous permeability. Material A was the first analyzed material (figure 4). Because the values of the inclusions relative permeability are not very high and the difference between the relative permeability of the matrix and inclusions is also small, it is expected that the two formulas are efficient in the analysis.

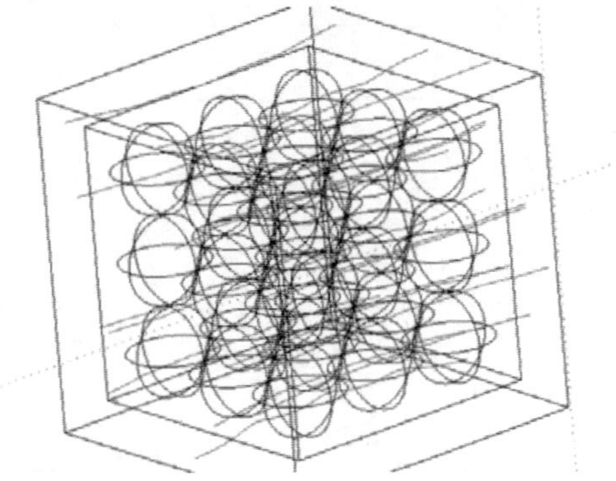

Figure 4. Magnetic field line plot for a model with 50% particle concentration

The magnetization was computed for all three cases (particle concentration is 9%, 30%, respectively 50%) and for five different values of the applied magnetic field (H_1 = 23.86 kA/m, H_2 = 47.81 kA/m, H_3 = 71.66 kA/m, H_4 = 119.39 kA/m, H_5 = 453.65 kA/m – see figure 1). Figure 5 is presents the variation of the magnetization for three different filler factors f = 0.09, f = 0.3 and f = 0.5, for μ_{r_P3} = 2.24 and B=0.3T. It is easy to observe that the magnetization values are approximately the same regardless of the homogenization formula.

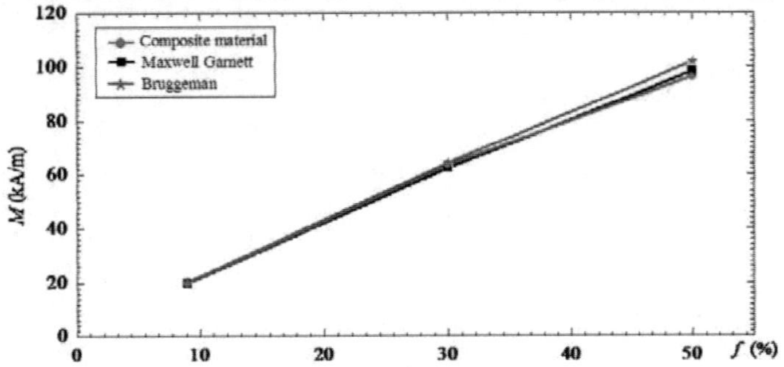

Figure 5. Magnetization variation for μ_{r_P3} = 2.24 and for three different concentrations (9%, 30%, 50%)

Journal of Physics: Conference Series **585** (2015) 012003 doi:10.1088/1742-6596/585/1/012003

In order to point better the difference between the obtained results, the relative error was calculated.

$$err = \left| \frac{M_{composite} - M_{homogeneous}}{M_{composite}} \right| \cdot 100\ [\%] \qquad (3)$$

Table 1 presents the relative errors between the computed composite material magnetization and the homogenized one. It can be observed that the errors are bigger with the increase of the difference between the two permeability values. Also, the errors obtained using Bruggeman formula are bigger than the ones using Maxwell Garnett.

Table 1. Relative error between the magnetization values

	f=0.09		f= 0.3		f= 0.5	
	M. G.	Brugg.	M. G.	Brugg.	M. G.	Brugg.
P1	2.17%	6.56%	2.73%	7.11%	0.78%	11.23%
P2	1.87%	4.98%	2.31%	4.67%	1.34%	8.75%
P3	1.25%	2.66%	1.79%	1.37%	2.28%	5.81%
P4	0.77%	1.05%	1.62%	0.95%	2.92%	3.69%
P5	0.85%	0.85%	1.29%	1.29%	3.09%	3.09%

There are two possible explications for these errors: one is the fact that if the value of f is relatively high (0.5 in this case) errors appear because the interactions between inclusions are high and the second one is the fact that performed analysis is made at a microscopic scale which brings some complications because the physical laws change at this scale.

However the differences between the values are not very high so, for this kind of composite materials, the two homogeneous formulas can be used in a numerical simulation of the real material behavior, for simple and also complex structures. The homogeneous results can approximate in acceptable limits the real results.

The second material that is analyzed in this paper is material B. Like in the previous case the material magnetization was computed for the composite material and for the homogeneous one. Three points were chosen: P1 – H=89.41 A/m, μ_{r_P1} = 6225.13; P2 – H=2161.89 A/m, μ_{r_P2} = 589.65; P3 – H=15296.81 A/m, μ_{r_P3} = 99.76 from the $\boldsymbol{B}$=f($\boldsymbol{H}$) magnetic characteristic obtained using the VSM device (figure 6).

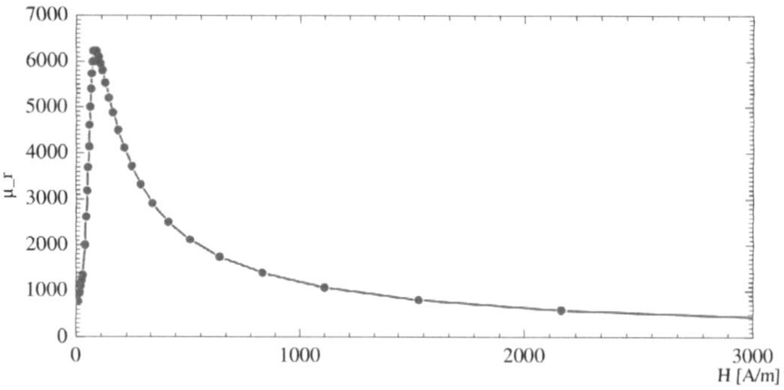

Figure 6. Magnetic permeability graph for material B

Like in the previous case an homogeneous field was applied in the x direction. Figure 7 presents the magnetic flux density plot for point P3 and for f = 0.3. Even thought the imposed field is homogeneous, one can observe that, the presence of inclusions disturbs the flux density values inside the plot.

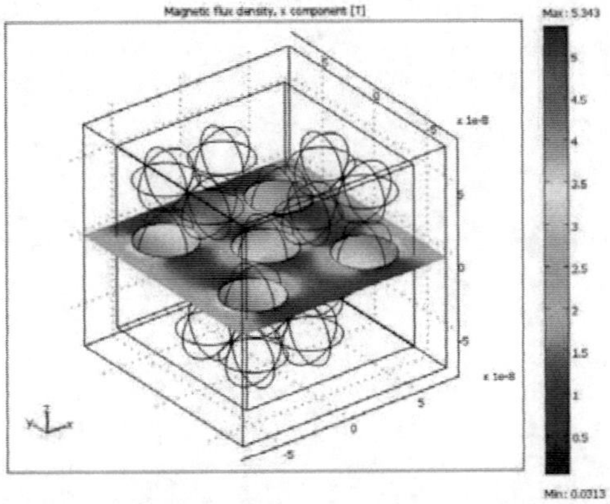

Figure 7. Magnetic flux density plot for a model with 30% particle concentration

For each permeability value, five different inclusion concentrations were considered - 50%, 40%, 30%, 20% and 9%. In the next three figures the magnetization variation versus the concentrations for each permeability value in part are presented.

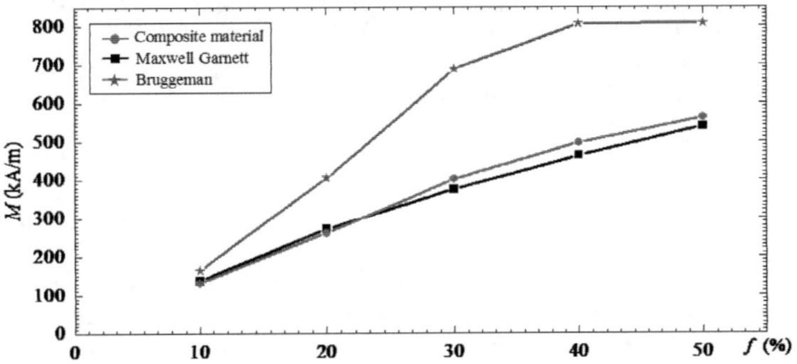

Figure 8. Magnetization variation for μ_{r_P1} versus different inclusions concentrations

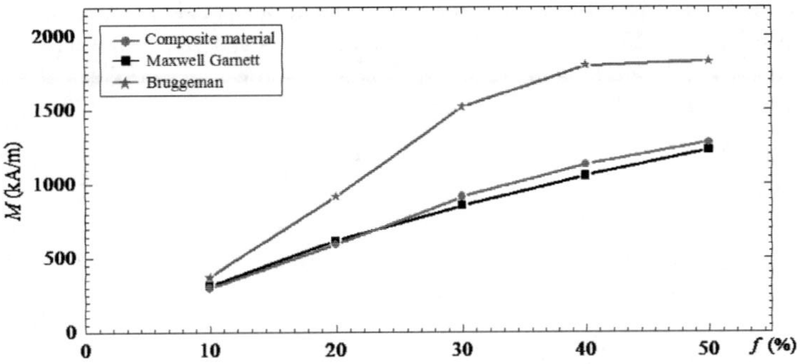

Figure 9. Magnetization variation for μ_{r_P2} versus different inclusions concentrations

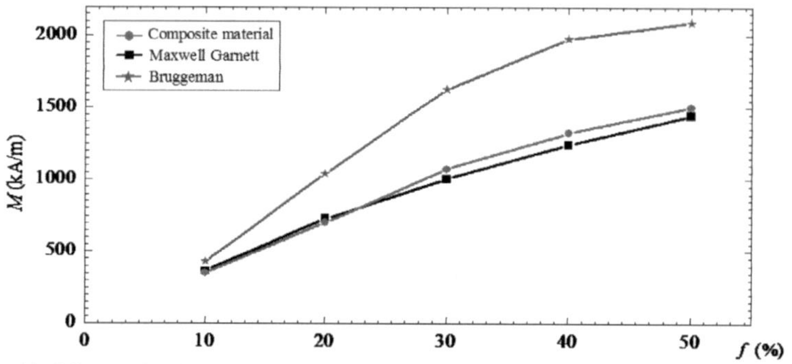

Figure 10. Magnetization variation for μ_{r_P3} versus different inclusions concentrations

It is easy to observe that the Maxwell Garnett formula is more efficient for modeling the composite material. The relative error is less than 5 % between the obtained results with Maxwell Garnett and the real material. Also, from the three figures one can easily observe that the Maxwell Garnett formula gives smaller results than Bruggeman. The difference is due to the fact that Maxwell Garnett does not take into consideration the inclusions interaction. In this case, because the configuration is not very complex, the impact of the fact that the local material behavior is not taken into account is not very big. But the impact of the concentration value is significant, the two formulas having a higher error for a bigger f value - the errors increase approximately two times. The errors for Bruggeman formula are ten times higher than the ones for Maxwell Garnett. Another factor that influences the errors is the fact that in this second case the difference between the permeability of the matrix and the inclusions is bigger than in the first case. This fact perturbs the accuracy of the results obtained using Bruggeman formula. In conclusion, for this case the Maxwell Garnett formula is also more efficient in showing the sample magnetic behavior.

5. Conclusion

From the simulations that were carried out one can observe that the results obtained using the Maxwell Garnett homogenization formula are closer to the ones obtained for the composite material. Although both formulas produce good results for many configurations and are successfully used in order to confirm the expected results in some cases, they are not so precise if the inclusion concentrations are too high or the configuration is too complex. It is important for each desired case to take into consideration the problem hypothesis (inclusions concentration, difference between matrix and inclusions magnetic permeability etc.) in order to be able to make some predictions on the formula accuracy for that specific case. In conclusion, for this analysis, the Maxwell Garnett formula was more efficient for the two considered materials than Bruggeman formula.

References

[1] Waki H, Igarashi H and Honma T 2005, *IEEE Trans. Magn.*, **41**, 1520 – 23.
[2] Orive G, Hernandez R M, Rodriguez Gascon A, Dominguez-Gil A and Pedraz J L 2003, *Current Opinion in Biotechnology*, **14**, 659-664.
[3] Pouponneau P, Leroux J C and Martel S 2009, *Biomaterials*, **30**, 6327 – 32.
[4] Chebak A, Viarouge P and Cros J 2009, *IEEE Trans. Magn.*, **45**, 952-955.
[5]....Ionita V, Petrescu L, Bordianu A, Tabara O, 2013 *Adv. Electr. Comp. Eng.* **13** (2), 121-126.
[6] Fu L, Macedo P B and Resca L. 1993, *Phys Rev B Condens Matter*, **47**, 13818 – 29.
[7] McPhedran R C and McKenzie D R 1978, *Proc R Soc Lond A*, **359**, 45 - 63.
[8] White K W 2000, *J Appl Phys*, **88**, 1962 - 70
[9] Sareni B, Krahenbuhl L, Beroual A and Brosseau C 1997, *J Appl Phys*, **81**, 2375 - 84
[10] Krakovsky I and Myroshnychenko V 2002, *J Appl Phys,* **92**, 6743 - 48

[11] Santugini-Repiquet K 2007, *J. Math. Anal. Appl.*, **334**, 502–516
[12] Banks H T, Bokil V A, Cioranescu D, Gibson N L, Grisco G and Miara B 2005 *Homogenization of Periodically Varying Coefficients in Electromagnetic Materials*, Technical Report #2005-2, Statistical and Applied Mathematical Sciences Institute
[13] Lakhtakia A and Weiglhofer W S 1998, *Int. J. Electronics*, **84**, 285 - 294
[14] Lakhtakia A and Weiglhofer W S 2000, *Int. J. Electronics*, **87**, 1401 - 08.
[15] Maxwell Garnett J C 1904, *Phil. Trans. R. Soc. Lond. A*, **203**, 385-420
[16] Sihvola A 1999 *Electromagnetic mixing formulas and applications*, The Institution of Electrical Engineers, (London, United Kingdom)
[17] *** COMSOL Multiphysics 3.4

Analysis of hysteretic spin transition and size effect in 3D spin crossover compounds investigated by Monte Carlo Entropic sampling technique in the framework of the Ising-type model

D Chiruta[1,2,3], **J Linares**[1], **P R Dahoo**[4] **and M Dimian**[3]

[1]GEMAC, *Université de Versailles Saint-Quentin-en-Yvelines*, CNRS-UVSQ (UMR 8635), 78035 Versailles Cedex, France

[2]LISV, *Université de Versailles Saint-Quentin-en-Yvelines*, 78140 Velizy, France

[3]Department of Electrical Engineering and Computer Science, *Stefan cel Mare University*, Suceava, 720229, Romania

[4]LATMOS, *Université de Versailles-Saint-Quentin-en-Yvelines*, CNRS-UPMC-UVSQ (UMR 8190), 78280 Guyancourt, France

E-mail: jorge.linares@uvsq.fr, dimian@eed.usv.ro

Abstract. In spin crossover (SCO) systems, the shape of the hysteresis curves are closely related to the interactions between the molecules, which these play an important role in the response of the system to an external parameter. The effects of short-range interactions on the different shape of the spin transition phenomena were investigated. In this contribution we solve the corresponding Hamiltonian for a three-dimensional SCO system taking into account short-range and long-range interaction using a biased Monte Carlo entropic sampling technique and a semi-analytical method. We discuss the competition between the two interactions which governs the low spin (LS) - high spin (HS) process for a three-dimensional network and the cooperative effects. We demonstrate a strong correlation between the shape of the transition and the strength of short-range interaction between molecules and we identified the role of the size for SCO systems.

1. Introduction

The predictions which consider that, the density of the future recording materials is approaching its limit, imply that more theoretical studies to be done to overcome this obstacle. In this respect, a close correlation between theoretical and experimental work must be established. A better understanding of the theoretical aspects, of the parameters in the network structure will lead synthetic chemistry towards elaborations of new spin crossover (SCO) materials with these different properties.In recent years the SCO phenomenon [1,2,3,4] has attracted a growing and important interest for physicists, biochemists, and chemists, since it offers an understanding of ligand field theory and the functions of molecular systems. The benefits for industrial applications are beneficing for data storage, fast sensors, memory devices or biological imaging.

The (Fe^{II}) SCO compounds contain bistable molecules which can exist in two different states: a diamagnetic low-spin state (LS) and a paramagnetic high-spin state (HS). Depending on certain factors (like temperature, pressure, magnetic field, ultra-short laser pulse, X-ray radiations, electrical field [5]) the states can switch. As a response to these factors, the LS-HS or HS-LS transition involves both structural changes and electronic transformations. Regarding the shapes of the spin transition, the Fe^{II} SCO compounds can present a gradual, abrupt, with hysteresis, or into two-step shape. The last two shapes, which results when the system confers a bistability as a function of some external factors (usually temperature), represent one of the most pregnant aspect of spin crossover phenomenon.

Nowadays SCO compounds have been studied intensively, and many studies and topics have present the influence of some external stimuli as temperature, magnetic field, pressure, light on the SCO behavior. Understanding and controlling the properties of bistable complexes, also involve studying the influence of the cooperative interaction between the molecules. It is well known [1] that hysteresis in a spin transition curve has two causes. The first one came from the phase change in the lattice and the second one come from the elastic interaction between neighbouring molecules in the network. Taking into account the second hypothesis, in this paper we study a spin transition accompanied by a hysteresis loop and we show the importance of the strength of short-range interaction in understanding the properties and predicting the behavior of SCO compounds [6, 7].

In our study we use the variation in temperature (one of the most common perturbation factor which influences the change spin state process) and an Ising-like model. On the basis of good agreement between simulated results and experimental ones, we considerate in the Ising Hamiltonian both short range and long range interactions as well as distinct degeneracies between the molecular states [6].

The article is organized as follows: in section 2, we present and explain the Ising-like model with both short range (J) and long range (G) interactions. In section 3 we present how the partition function formula with the contribution of Monte Carlo (MC) entropic sampling method is realized. We also present the role of the short-range interactions and an analysis of the system behavior as a function of several physical parameters. In section 4 we present how the number of molecules influences the cooperative phenomena. The last section ends with conclusions.

2. The model

In recent years a variety of models (Bragg-Williams models [8], Ising models [9, 10], atom-phonon (APC) model [11, 12, 13, 14, 15, 16, 17, 18, 19] or mechano-elastic model [20, 21]) and studies have been performed to explain the LS-HS transitions, which appear in some molecular solids. It is well known that these switching processes appear in the SCO compounds which contain metal ions with $d^4 - d^7$ electron configuration in an octahedral symmetry environment. According to how the electrons occupy and move between the e_g and t_{2g} oribitals, the two LS and HS states, for Fe^{II}, can be defined as follows:

- The Fe^{II} is in the LS state when all the six electrons occupy the three orbitals of the lowest energy. In this case, the spin paring energy (∏) is less than the ligand field splitting energy (δ).
- The Fe^{II} is in the HS state when five electrons occupy all the five orbitals and sixth electron will be in of the lowest energy orbital. In this case the ligand field splitting energy(δ) is less than spin paring energy (∏)[figure 1]

The thermodynamic competition of these two states (LS and HS) can be characterized by the proportion of molecules in the HS state, denoted here as n_{HS}.

We would like now to explore further this n_{HS} distribution as a function of the temperature. To do this, we will adopt a two level model (Ising model) in which we consider a three dimensional network with free boundary conditions. Figure 2 shows a random configuration for 3x3x3 (27 molecules) system.

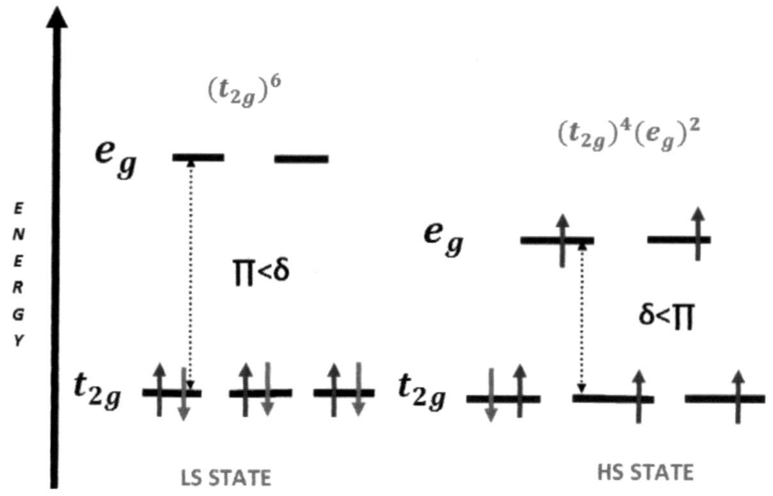

Figure 1. Distribution of electrons over 3d orbitals in low-spin state and high-spin state, for Fe^{II} .

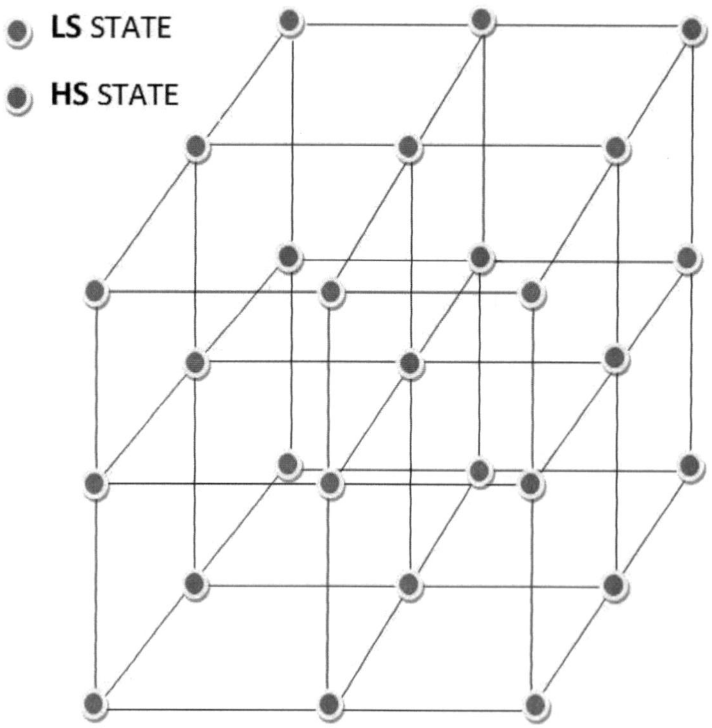

Figure 2. A three-dimensional (3x3x3) network containing twenty-seven molecules chosen arbitrarily.

Taking into account that: (1) to each molecule we can associate a fictitious spin operator "σ" with two eigenvalues (-1 for LS state and +1 for HS state) ;(2) Δ is the energy gap between the HS state and LS state; (3) $g_{state} = \left(\frac{g_{HS}}{g_{LS}}\right)$ (g_{HS} is the degeneracy of the HS state and g_{LS} is the degeneracy of the LS state); and (4) J is the short-range interactions and G represent the long-range interactions the Ising Hamiltonian can be written as[6]:

$$H = -h_f \sum_{i=1}^{N} \sigma_i - J \sum_{\langle i,j \rangle} \sigma_i \sigma_j \quad (1a)$$

where:

$$h_f = -\left(\frac{\Delta - k_B T \ln(g_{state})}{2} - G\langle\sigma\rangle\right) \quad (1b)$$

- k_B is Boltzmann's constant and T is the temperature of the system.

3. Analysis of the method and short-range interaction effect

In this work we adopt a consistent method based on the bisection method [22] and we show the evolution of n_{HS}, high-spin fraction, as a function of temperature for different physical parameters values.

Using equation 1b and taking into account that n_{HS} = (1+σ)/2, we can define the high-spin fraction corresponding to the Ising Hamitonian as:

$$n^1{}_{HS} = \frac{1 + \frac{h_f}{G} + \frac{\Delta - k_B T \ln(g_{state})}{2G}}{2} \quad (2)$$

If we consider a system of equations, we will need now a second function $n^2{}_{HS}$ to find the solutions. The $n^2{}_{HS}$ fraction can be found from the MC entropic sampling method [23, 24]. This method has been used by Linares et al.[22] who showed that, by comparison to exact solution for a one dimensional network (1D systems), that this MC entropic sampling method gives the same results.
In this article we apply this technique to a three dimensional network and we show that the hysteresis loop width is correlated to the short-range interaction strength.

Taking into account that d(m,s) is actually the degeneracy of the macrostate (m, s) where m and s are two macroscopic variables $m = \sum_{j=1,N} \sigma_j$; $s = \sum_{<i,j>} \sigma_i \sigma_j$ the expression for $n^2{}_{HS}$ is :

$$n^2{}_{HS} = \frac{\sum_{j=1,N} \frac{m_j}{N} d(m_j s_j) \exp(-\beta(-h_f m_j - J s_j))}{\sum_{j=1,N} d(m_j s_j) \exp(-\beta(-h_f m_j - J s_j))} \quad (3)$$

where $\beta = 1/k_B T$;

Taking into account the expressions of $n^1{}_{HS}$ and $n^2{}_{HS}$ we can construct the graph for each temperature.

Figure 3 displays the n_{HS} as a function of effective field h_f (equation 1b). The graphic for temperature T=110K is represented in Figure (3a). In this case because the two curves ($n^1{}_{HS}$ and $n^2{}_{HS}$) intersect at only one point, this will be the solution. In this case the intersection point (point A) is a stable point and can also be indentified in figure 4.

Now for other temperatures we can also have one solution (point E) for T=240K or three-solutions (points B, C, D) for T=160K. For T=240K (figure 3b) it can be observed that we have also a stable point (point E) but with another n_{HS} associates [for point A the n_{HS} is equal to 0. and for point E the corresponding n_{HS} is equal to 0.99]. When three solutions are obtained, as for T=160 K (figure 3c) we will have:

- point B – stable solution with $n_{HS} = 0$.
- point C – unstable solution with $n_{HS} = 0.72$
- point D – metastable solution with $n_{HS} = 0.99$

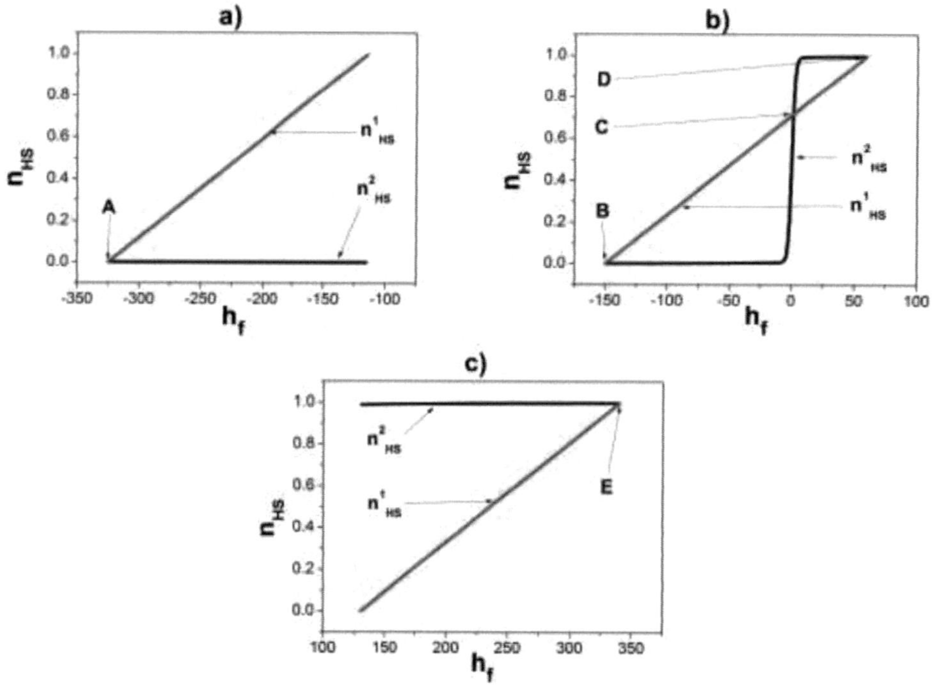

Figure 3. Evolution of HS fraction n_{HS} for a 3D system (64 molecules) with open boundary conditions for T=110K (a) T=160K (b) T=240K (c). The parameter values are Δ= 1209K, ln (g_{state})= 7, long-range interaction is *G*=105K and the short-range interaction is *J*= 100K.

Figures 4, 5, 6, 8 displays the evolution of HS fraction n_{HS} as a function of temperature and are made in the manner already described.
To study the behavior of SCO complex, it is considered that: T_{up} is the ascending thermal transition temperature, T_{down} is the descending thermal transition temperature and $T_{1/2}$ is the average temperature between T_{up} and T_{down}. For a short–range interaction (J) equal to 100K and a long-range interaction (G) equal to 105K, then figure 4 is obtained. In this figure if we consider that ΔT is the hysteresis width and represent the difference between the ascending thermal transition temperature (T_{up}) and the descending thermal transition temperature (T_{down}), then a hysteresis of ΔT=53K is observed which

corresponds to T_{up}=198K and T_{down}=145K. Moreover in this case when the system is formed by 64 molecules it can be observed that $T_{1/2}$ is equal to 171.5K.

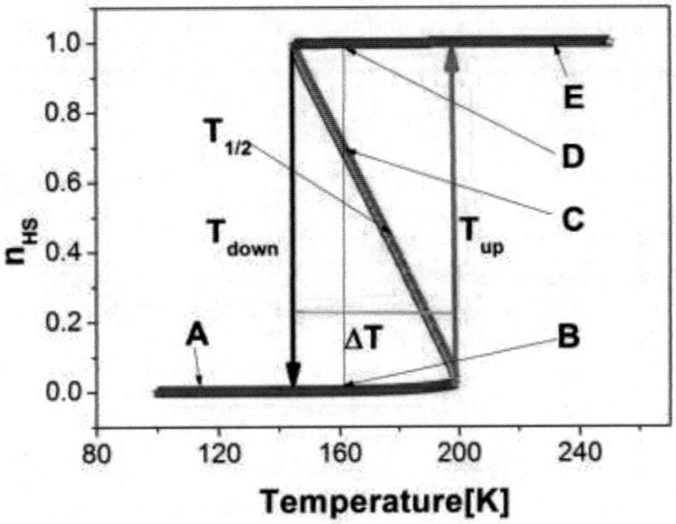

Figure 4. Evolution of HS fraction n_{HS} as function of temperature, for a 3D system (64 molecules) with open boundary conditions. The parameter values are Δ= 1209K, ln(g_{state})= 7, long-range interaction is *G*=105K and the short-range interaction is *J*= 100K.

When the short-range interactions are decreased to J=30K (figure 5) we report also a decrease in the hysteresis loop dimension from ΔT=53K to ΔT=8K. In this case even if the temperature $T_{1/2}$ remains the same as in the previous example (figure 4) the ascending temperature T_{up} and the descending temperature T_{down} are changing. Now T_{up} is equal to 175K, T_{down} is equal 167K and the hysteresis width ΔT = 8 K.

When the short–range interaction is decreasing futher, to J=10K(figure 6), the interactions are not strong enough to produce a hysteretic behavior and the transition shape now have a new shape compared to that of figure 4.

It can be seen in figure 4 , when the system presents a strong short-range interaction equal to 100K the shape of the transition is a “Z” shape and when the interaction between the molecules is decresing J=30K then the transition resembles a called “S shape” in some article (see figure 5).

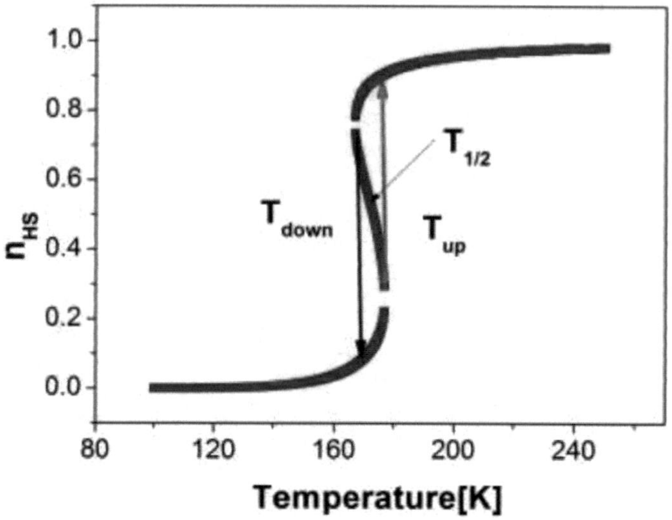

Figure 5. Evolution of HS fraction n_{HS} as function of temperature, for a 3D system (64 molecules) with open boundary conditions. The parameter values are Δ= 1209K, ln(g_{state})= 7, long-range interaction is *G*=105K and the short-range interaction is *J*= 30K.

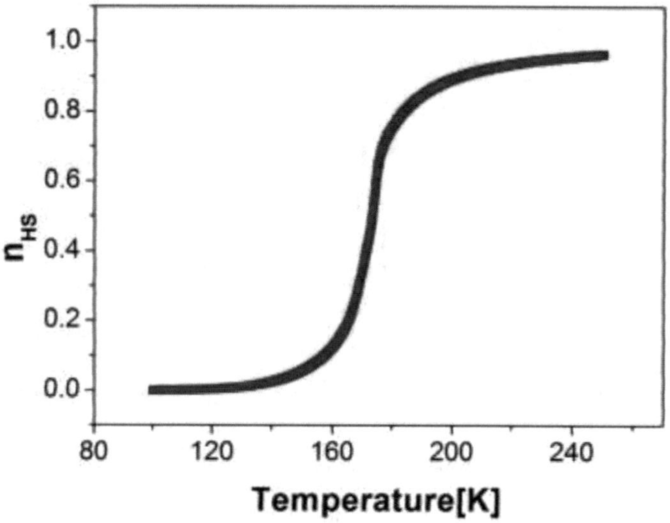

Figure 6. Evolution of HS fraction n_{HS} as function of temperature, for a 3D system (64 molecules) with open boundary conditions. The parameter values are Δ= 1209K, ln(g_{state})= 7, long-range interaction is *G*=105K and the short-range interaction is *J*= 10K.

The hysteresis loop width for different short-range interactions is presented in figure 7. In this case, the system will have a hysteresis width ΔT equal to 53 K corresponding to T_{up} = 198 K and T_{down}= 145

K, for long-range interactions G = 105K [X point]. When these interactions decrease to 10K (J = 10 K), the hysteresis loop width - ΔT will be equal to 0 K [Y point].

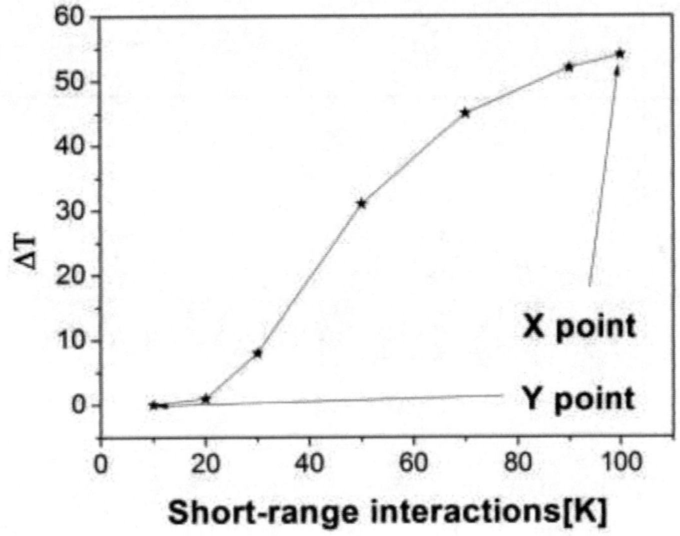

Figure 7. Evolution of hysteresis loop width ΔT vs. short-range interaction parameter J. The parameter values are Δ= 1209K, ln(g_{state})= 7, long-range interaction is *G*=105K.

4. Size effect

A similar behavior [figure 8] has been noticed, when the intensity of the short-range interactions is changing for a one or two dimensional network [22]. In this contribution we also fix the interactions and we change only the number of molecules N. Here, when the number of molecules of the system is increasing, the interactions will be stronger and a S shape will appear. For another set of parameters, for example, if is increased the short-range interactions [$J_1 = x * J$ (with x take values from 0 to ∞, J=1K)], this "S shape" will transform to a "Z-shape" as in figure 4.

Figure 9 presents the evolution of the hysteresis loop width for different sizes and short-range interactions. The hysteresis width increases, with the number of molecules, and as a function of the strength of short-range interactions for N=216 molecules we have different values. When between the molecules is considerate a short-range interactions J=17K we can observe for N=8 and N=27 molecules, that, the system don't present hysteresis. For the same short-range interactions system present a hysteresis loop equal to 1K when the number of the molecules is N= 216 molecules.
Moreover when the short-range interactions is increased to J=22K for N=8 molecules there will be no hysteresis but for N=27 molecules there will be a hysteresis loop equal with 1K. In this case it can be observed, the fact that, a system with N=27 molecules and a short-range interaction between molecules of J=22K, have the same hysteresis loop dimension as a system with N=216 molecules but with a less short-range interaction (J=17K).

Concerning the number of molecules (size effect) for a system, other studies have been performed by Rotaru et al. [19,26] using the FORC technique and by Murukoa et al.[27] taking into account the edge effect. In all cases it was shown that number of molecules (system size) influence the hysteresis loop dimension.

Journal of Physics: Conference Series **585** (2015) 012004 doi:10.1088/1742-6596/585/1/012004

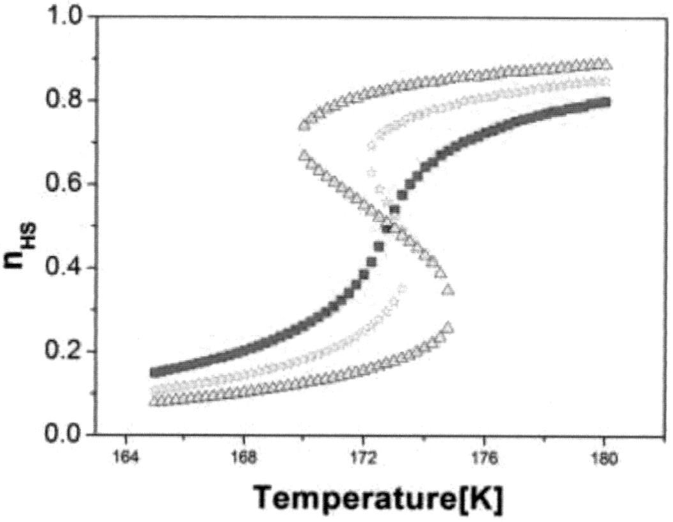

Figure 8. Evolution of HS fraction n_{HS} as function of temperature, for a 3D system with different number of molecules N=216(triangle), N=64(stars) N=8(square) with open boundary conditions. The parameter values are Δ= 1209K, ln(g_{state})= 7, long-range interaction is *G*=105K and the short-range interaction is *J*= 22K.

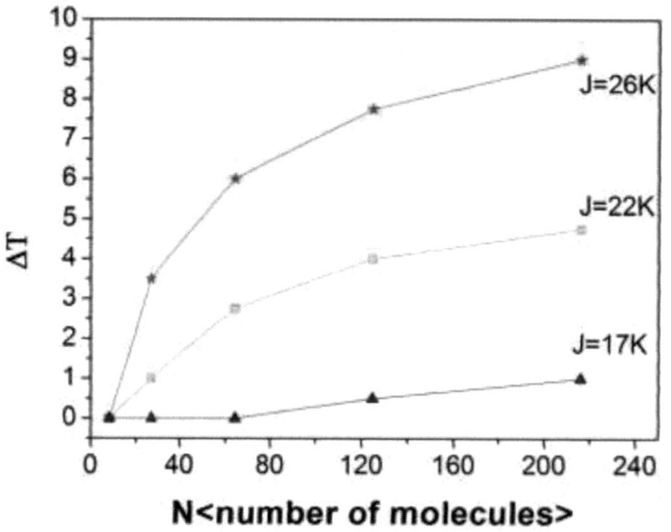

Figure 9. Evolution of hysteresys loop width ΔT vs. number of molecules for different short-range interaction strenght. J=17K(triangle), J=22K(square) J=26(stars). The parameter values are Δ= 1209K, ln(g_{state})= 7, long range interaction is *G*=105K.

5. Conclusions

We have used Ising-type model with short-range and long-range interactions to investigate the influence of the short-range interactions and of the system size on a three-dimensional system. Applying MC entropic sampling technique to a 3D SCO complex proves to be a good way to characterize the hysteretic spin transition behavior.

Another result obtained is the changing of hysteresis curves from a Z shape to a sinusoidal(S) shape when the short-range interactions are decreasing. This behavior has been also obtained by Linares et al.[6] for one-dimensional $[Fe(Htrz)_2(trz)(BFA)_2]$ compound or two-dimensional $[Fe(btr)_2(NCS)_2H_2O]$ compound. And finally we have found and confirm the hypothesis [25, 26] that, *the hysteresis width decrease as of number of molecules is reduced.*

6. Acknowledgements

This paper was supported by the Conseil regional d'Ile-de-France and by the Romanian National Program II, no. 107/06.08.2010. M.D. thanks the project "Progress and development through postdoctoral research and innovation in engineering and applied sciences - PRiDE - Contract no. POSDRU/89/1.5/S/57083", project co-funded from European Social Fund through Sectorial Operational Program Human Resources 2007-2013.

7. References

[1] Gutlich P and Goodwin H E 2004 Editors, Spin Crossover in Transition Metal Compounds I II and III *Topics in Current Chemistry* **233 234 235** Springer-Verlag, Berlin, Heidelberg, New York.
[2] Spieging H, Kohlaas T, Romsted H, Hauser A, Bruns-Yilmaz C, Kusz J and Gütlich P 1999 *Coord. Chem. Rev.* **192** 629.
[3] Linares J, Codjovi E and Garcia Y 2012 *Sensors* **12** 4.
[4] Kröber J, Audière J P, Claude R, Codjovi E, Kahn O, Haasnot J G, Grolière F, Jay F, Boussekou A, Linares J, Varret F, and Gonthier-Vassal A 1994 *Chem. Mater.* **6** 1404.
[5] Rotaru A, Gural'skiy I y A, Molnar G, Salmon L, Demont P and Bousseksou A 2012 *Chemical Communications* **48** 4163.
[6] Linares J, Spiering H and Varret F 1999 *The European Physical Journal B* **10** 271.
[7] Boukheddaden K, Linares J, Spiering H and Varret F 2000 *The European Physical Journal B* **15** 317-326.
[8] Koudriavtsev A.B. 1999 *Chemical Physics* **241** 109-126.
[9] Chiruta D, Linares J, Dahoo P R and Dimian M 2012 *Journal of Applied Physics* **112** 074906.
[10] Boussekou A, Nasser J, Linares J, Boukheddaden K and Varret F 1992 *J. Physique I* **2** 1381.
[11] Rotaru A, Carmona A, Combaud F, Linares J, Stancu A and Nasser J 2009 *Polyhedron* **28** 1684.
[12] Nasser J A 2001 *Eur. Phys. J. B* **21** 3.
[13] Nasser J, Boukheddaden K and Linares J 2004 *Eur. Phys. J. B* **39** 219.
[14] Rotaru A and Linares J 2007 *J. Optoelectronics. Adv. Mater.* **9** 2724.
[15] Gindulescu A, Rotaru A, Linares J, Dimian M and Nasser J 2010 *J. Appl. Phys.* **107** 09A959.
[16] Gîndulescu A, Rotaru A, Linares J, Dimian M and Nasser J 2011 *Polyhedron* **30** 3186.
[17] Rotaru A, Linares J, Codjovi E, Nasser J and Stancu A 2008 *J. Appl. Phys.* **103** 07B908.
[18] Rotaru A, Linares J, Mordelet S, Stancu A and Nasser J 2009 *J. Appl. Phys.* **106** 043507.
[19] Rotaru A, Rotaru G-M, Linares J and Garcia Y 2012 *J. Optoelectronics. Adv. Mater* **14** 529.
[20] Enachescu C, Stoleriu L, Stancu A, Hauser A 2011 *Journal of Applied Physics* **109** 07B111.
[21] Enachescu C, Nishino M, Miyashita S, Hauser A, Stancu A and Stoleriu L 2010 *Europhys. Lett.* **91** 27003.
[22] Linares J, Enachescu C, Boukheddaden K and Varret F 2003 *Polyhedron* **22** 2453.

[23] Lee J 1993 *Phys. Rev. Lett.* **71** 211.
[24] Shteto I, Linares J and Varret F 1997 *Phys. Rev. E* **56** 5128.
[25] Rotaru A, Varret F, Gindulescu A, Linares J, Stancu A, Létard J, Forestier T and Etrillard J 2011 *The European Physical Journal B - Condensed Matter and Complex Systems* **84** 439.
[26] Murukoa A, Boukheddaden K, Linares J and Varret F 2011 *Phys. Rev. B* **84** 054119.

Hysteresis in column systems

P Ivanyi and A Ivanyi
Pollack Mihaly Faculty of Engineering and Information Technology, University of Pecs H-7624 Hungary

E-mail: peteri@morpheus.pte.hu

Abstract. In this paper one column of a telescopic construction of a bell tower is investigated. The hinges at the support of the column and at the connecting joint between the upper and lower columns are modelled with rotational springs. The characteristics of the springs are assumed to be non-linear and the hysteresis property of them is represented with the Preisach hysteresis model. The mass of the columns and the bell with the fly are concentrated to the top of the column. The tolling process is simulated with a cycling load. The elements of the column are considered completely rigid. The time iteration of the non-linear equations of the motion is evaluated by the Crank-Nicolson schema and the implemented non-linear hysteresis is handled by the fix-point technique. The numerical simulation of the dynamic system is carried out under different combination of soft, medium and hard hysteresis properties of hinges.

1. Introduction

Several engineering problems deal with the modelling of the non-linear behaviour of systems. The non-linear property of the engineering structures can be handled by single-valued or two-valued hysteretic characteristics. Generally two questions arise here, how to model the non-linear, hysteretic characteristics of the structure and how to insert this model into the simulation method of the system.

The non-linearity of thermal, magnetic and mechanical systems can be realized with microscopic, mesoscopic and macroscopic simulations. The simplest representation of non-linear characteristics is the macroscopic model, where the purpose is the phenomenological description of the physical process. On the other hand a more detailed approach to describe the non-linear behaviour of the structure or material is the microscopic model, where the elementary scale of the material is simulated from the energetic point of view. The non-linearity of material domains/clusters can be analysed by the Stoner-Wohlfarth type [1] and the Jiles-Atherton type models [2]. The mesoscopic description of the hysteresis behaviour with respect to some physical properties can be found in the Preisach type models [3]-[10], which is the most popular modelling of non-linearity of materials and structures.

The above-summarized general trend can also be found in the simulation of non-linear behaviour of mechanical systems.

To handle the non-linearity of joints in steel frames different phenomenological models have been developed. The range of these models varies from the so-called empirical models through the analytical simulations [11], [12] to the very popular different level of the Ramberg-Osgood and the Richard-Abbott models [13]-[19]. To build up the material from elementary clusters and modelling their mechanical behaviour is developed in microscopic models in Reference [20]. To represent the hysteresis of steel due to the stress and strain effects the modified versions of Jiles-Athertom models can be used [21], [22] and in the modified models with loss separation under different stresses [23].

There is a continued research to extend the Preisach model for the simulation of mechanical properties of materials and to describe the non-linear behaviour of steel structures [24]-[26].

The purpose of the recent research is to extend the Preisach model for the theoretical investigation of semi-rigid joints of hinges in a two storey steel column structure and insert the model into the computation of the column systems.

2. The bell tower

In Pécs, city of Hungary, there is a Turkish style mosque on the main square of the city centre, which functions as a Christian church nowadays. Pécs was the Cultural Capital of Europe in 2010. During the preparation period of this event a bell tower has been designed to the northeast corner of the mosque with the statue of St. Bartholomew (figure 1) by two architects: Zoltan Bachman and Balint Bachmann [27]. The three bells are surrounded by three slender steel columns. The bell tower has a moving telescopic structure, which can hydraulically rise to become a tower (figure 2). The designed form of the bell tower can be seen in figure 3.

Considering the structure of the bell tower, two sensitive points should be investigated during the tolling action, the fixed support at the bottom and the joint between the lower and upper columns. To model the behaviour of this dynamic system, first, one column with semi rigid hinges of non-linear hysteretic characteristic is modelled. In this way it is possible to check how the hinges behave during the dynamic action.

Figure 1. St Bartholomew's bell tower.

Figure 2. The telescopic height of the bell tower.

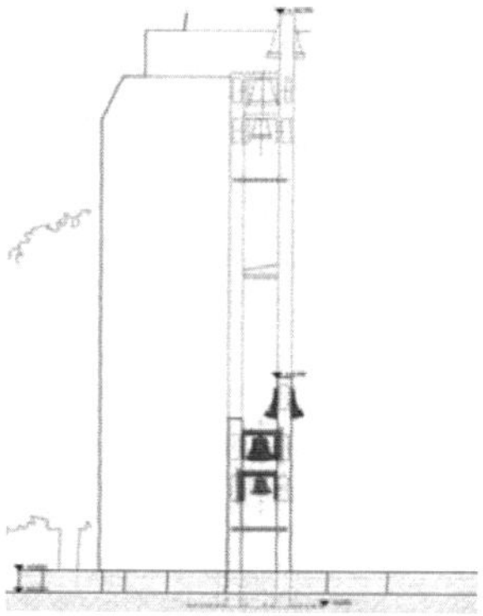

Figure 3. The designed form of the bell tower.

3. The model of the dynamic system

In this paper one column of the bell tower is modelled. It is built from two cylindrical shape tubes with the lower part of diameter $d_{1ex} = 0.3$ m and the thickness is 8 mm ($d_{1in} = 0.284$ m). The upper part has diameters $d_{2ex} = 0.282$ m and $d_{2in} = 0.266$ m. The length of the lower column is $l_1 = 5.35$ m, while the upper column is $l_2 = 8$ m long. The masses m_1, m_2 of the columns are concentrated to the upper end of the columns. The mass of the bell (290 kg) and the fly (40 kg) (figure 4) is added to the mass of the upper column.

As a result of the cycling external force the columns have declinations angles φ_1, φ_2 around of the joint points, resulting in deflections u_1, u_2 (figure 5). Considering small declination $u \approx \varphi l$, ($\sin\varphi \approx \tan\varphi \approx \varphi$, $\cos\varphi \approx 1$) a bending moment from the external cycling force and from the mass of the columns arises.

Journal of Physics: Conference Series **585** (2015) 012005 doi:10.1088/1742-6596/585/1/012005

To model the moment of the electrical motor for forcing the bells to toll, a concentrated, horizontal, exponentially increasing, periodically changing force is acting during the tolling process with the amplitude $F_0 = 1\,\text{kN}$ (figure 6) and cycling periodicity $T_p = 0.2\,\text{s}$,

$$F(t) = F_0\left(1 - e^{-t/T_p}\right)\sin\left(\frac{2\pi}{T_p}t\right). \tag{1}$$

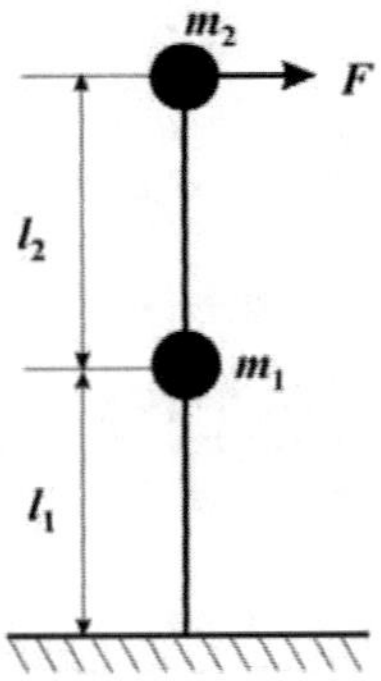

Figure 4. The column system.

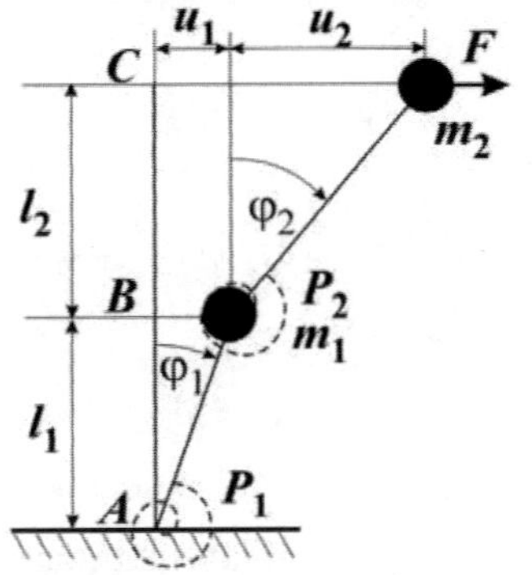

Figure 5. Deflection and declination of the columns.

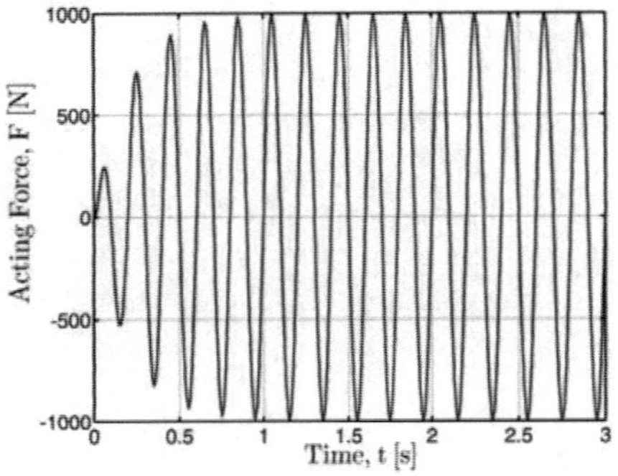

Figure 6. The cycling load of the columns.

The two parts of the column can be considered as a semi rigid cantilever. The bottom point of the lower column (point A) and the joining point between the two columns (point B) are modelled with rotational springs. The behaviour of the rotational springs is modelled with hysteresis characteristics between the spring moment (*P*) and the declination angle (φ), $\varphi = \mathcal{H}\{P\}$. During the investigations three types of characteristics are considered for the rotational springs as soft (H1), medium (H2) and hard (H3) as it can be seen in figure 7.

In mechanical systems the characteristics of spring moment-declination angle are used, however in this case the Preisach hysteresis model is introduced to describe the behaviour of the rotational springs, so the inverse hysteresis is constructed (figure 8).

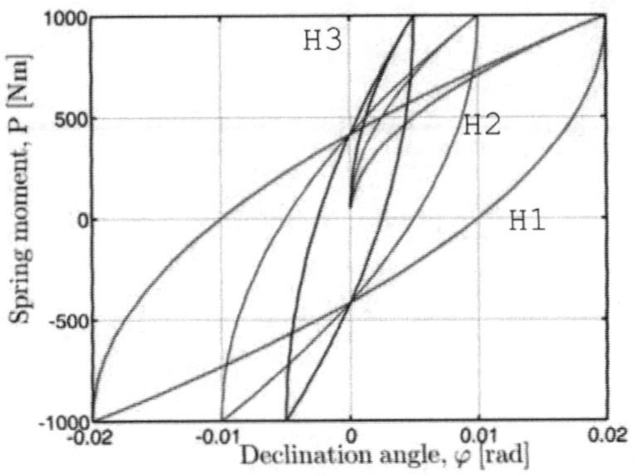

Figure 7. Direct characteristics of the rotational springs.

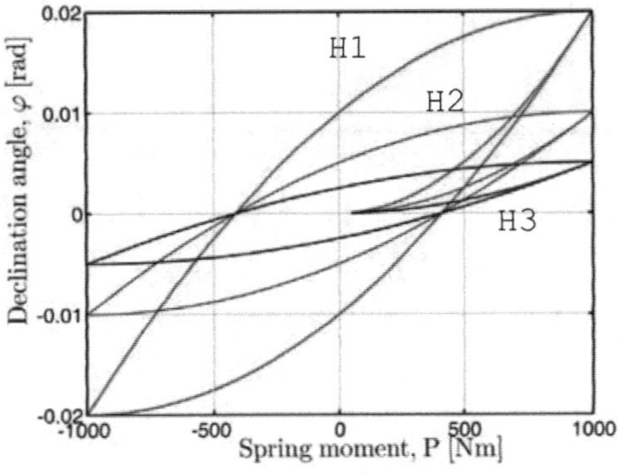

Figure 8. The Preisach model of hysteresis.

Introducing the elementary hysteron $\gamma(\alpha,\beta)$ of the Preisach model for handling the non-linear behaviour of the elementary cluster of the rotational spring, where α,β are the switching fields, the distribution function $\mu(\alpha,\beta)$ weights the effect of the elementary operator $\gamma(\alpha,\beta)$, while $P(t)$ is the acting spring moment [28]. The declination angle as the output of the non-linear spring can be determined as

$$\varphi(t,P)= \iint_{\alpha\geq\beta} \mu(\alpha,\beta)\gamma(\alpha,\beta)P(t)\,d\alpha\, d\beta . \tag{2}$$

The integration is evaluated over the Preisach triangle, defined as $\alpha \geq b$, while $\alpha \in [-1,\cdots,1]$, $\beta = [-1,\cdots,1]$.

4. Formulation of the dynamic motion

The elements of the steel columns are considered to be completely rigid. To describe the dynamic behaviour of the above system, the relation between the bending moment and the declination has to be determined. As the system plotted in figure 4 has two bars with two hinges at points (A) and (B), the motion will be determined by the bending moment acting at these points [29], [30]. The system damping effect is represented with the hysteresis loss.

The acting bending moment originating from the external force and the mass positioned to the top end of the columns accelerates the rotation and acts on the rotational spring at point (A) as

$$I_{2A}\ddot{\varphi}_2 + I_{1A}\ddot{\varphi}_1 + P_1 = F(l_1 + l_2) + G_2(u_1 + u_2) + G_1 u_1, \tag{3}$$

where I_{1A} and I_{2A} are the inertia moments of mass m_1 and m_2 at point A,

$$I_{1A} = m_1 l_1^2, \quad I_{2A} = m_2 (l_1 + l_2)^2, \tag{4}$$

P_1 is the bending moment action on the rotational spring at point (A) and G_1, G_2 are the gravity forces of m_1, m_2.

At point B the upper column acts and the acceleration of rotation can be formulated as

$$I_{2b}\ddot{\varphi}_2 + P_2 = F l_2 + G_2 u_2, \tag{5}$$

where I_{2B} is the inertia moment of mass m_2 at point B,

$$I_{2B} = m_2 l_2^2 . \tag{6}$$

Taking into account the condition of small declination and representing the deflections as

$$u_1 = \varphi_1 l_1, \quad u_2 = \varphi_2 l_2 , \tag{7}$$

the second order non-linear differential equations have the form of:

$$\begin{aligned} &I_{2A}\ddot{\varphi}_2 + I_{1A}\ddot{\varphi}_1 + P_1 = F(l_1 + l_2) + G_2[\varphi_1 l_1 + \varphi_2 l_2] + G_1 \varphi_1 l_1, \\ &I_{2b}\ddot{\varphi}_2 + P_2 = F l_2 + G_2 \varphi_2 l_2, \end{aligned} \tag{8}$$

$$\varphi_1 = \mathcal{H}\{P_1\}, \quad \varphi_2 = \mathcal{H}\{P_2\}. \tag{9}$$

5. Numerical iteration

An analytical solution of the above dynamic problem can be determined as it is proved in [31]. For the numerical approximation of the dynamic problem (8), (9) the time discretisation is evaluated by the double application of the Crank-Nicolson schema [32]. The non-linear iteration is realized by the fix-point technique [33], [34] to have a contract transformation of the direct characteristics

$$P(\varphi) = k_{FP}\varphi + R , \tag{10}$$

where k_{FP} is the fix-point constant to represent the linear part of the connection stiffness, $k_{FP}\varphi$ and R is the residual non-linearity. Substituting (10) into (8) at a fixed time moment, n,

$$\begin{aligned} I_{2A}\ddot{\varphi}_{2n}^{i+1} + I_{1A}\ddot{\varphi}_{1n}^{i+1} + k_{AFP}\varphi_{1n}^{i+1} &= M_{An}^{i} - R_{An}^{i}, \\ I_{2b}\ddot{\varphi}_{2n}^{i+1} + k_{BFP}\varphi_{2n}^{i+1} &= M_{Bn}^{i} - R_{Bn}^{i}, \end{aligned} \tag{11}$$

where M_{An}^{i} and M_{Bn}^{i} are the bending moments at points (A) and (B), at time moment n of iteration step i

$$\begin{aligned} M_{An}^{i} &= F_n(l_1 + l_2) + G_2\left(\varphi_{1n}^{i} l_1 + \varphi_{2n}^{i} l_2\right) + G_1\varphi_{1n}^{i} l_1, \\ M_{Bn}^{i} &= F_n l_2 + G_2\varphi_{2n}^{i} l_2. \end{aligned} \tag{12}$$

The iteration steps are as follows:

Step 1, At the new $n-$th time step the initial value of the declination angle is equal to the last value of the previous time step, $\varphi_n^i = \varphi_{n-1}$. The initial value of the non-linear part of the hysteresis is equal to the last value of the previous time step, $R_n^i = R_{n-1}$;

Step 2, The value of the acting bending moments are known from (12) with respect to the iteration of declination φ_n^i;

Step 3, With the solution of (11) a new iteration for the declination φ_n^{i+1} can be determined;

Step 4, An estimation for the bending moment acting on the rotational spring can be determined as $P_n^{i+1} \approx k_{FP}\varphi_n^{i+1} + R_n^i$;

Step 5, With the hysteresis (9) the remaining non-linear part can be calculated as $R_n^{i+1} = P_n^{i+1} - k_{FP} \cdot \mathcal{H}\{P_n^{i+1}\}$;

Step 6, The iteration continues on while $\left\|R_n^{i+1} - R_n^i\right\| > \varepsilon\left\|R_n^{i+1}\right\|$, otherwise $x_n^i = x_n^{i+1}$, $R_n^i = R_n^{i+1}$ and go to Step 2.

6. Simulation results

The numerical realization of the above theory has been developed in MATLAB code and the hysteresis has been introduced to the numerical iteration.

In all cases the time interval of the investigation is 3 s. The cycling time was selected as 0.2 s, in one period 20 time steps have been investigated, so the time discretisation was 0.01 s.

According to the different techniques to connect the elements of the telescopic construction and to fix the support of the bottom column, different properties of the rotational springs at points (A) and (B) have been considered. In table 1 the properties of the investigated rotational spring configurations are listed.

Table 1. Rotational springs at the nodes

Case	1	2	3	4	5
Point A	H2	H2	H1	H1	H2
Point B	H2	H3	H2	H3	H1

Case 1: First, it is assumed that both of the rotational springs have the same non-linearity, the hysteresis characteristics are assumed to be medium type (H2) at points (A) and (B).

At point (A) the variation of the declination angle versus the spring moment can be seen in figure 9, while at point (B) the hysteresis of the rotational spring generated between the declination angle and the spring moment is plotted in figure 10.

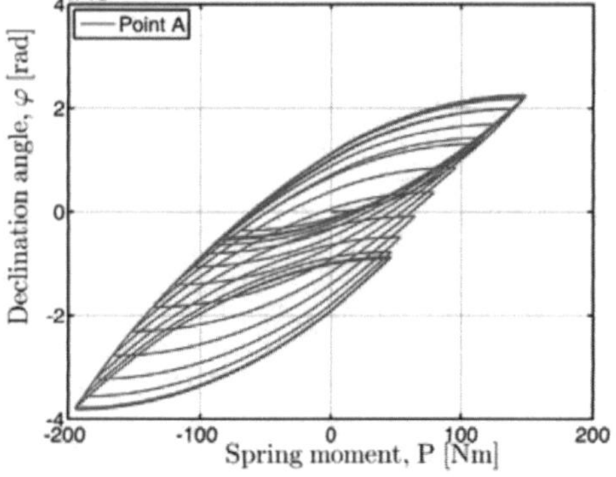

Figure 9. The hysteresis of the rotational spring at point (A).

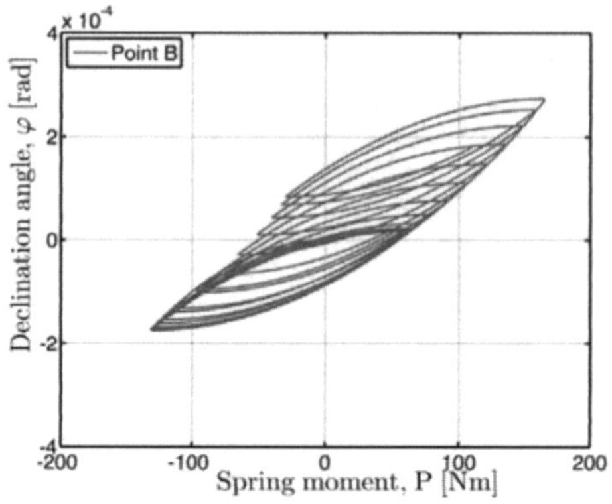

Figure 10. The hysteresis of the rotational spring at point (B).

From the figures it can be seen, that although both of the rotational spring has the same property, according to the different loads and positions, the two springs have different hysteresis. At the bottom point (A) of the lower column the rotation of the spring is more intensive compared to the spring at the joint between the upper and lower columns, point (B).

Similar properties are shown by the deflections of the end points of columns at points (B) and (C). According to the acting force the end points of the columns have deflection u_1 at the joint of the columns (B) and u_2 at the top point of the upper column (C), as it can be seen in figure 11 and figure 12.

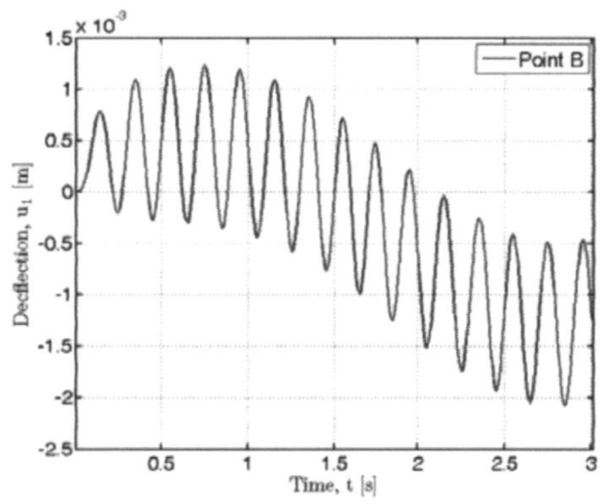

Figure 11. Deflection at point (B).

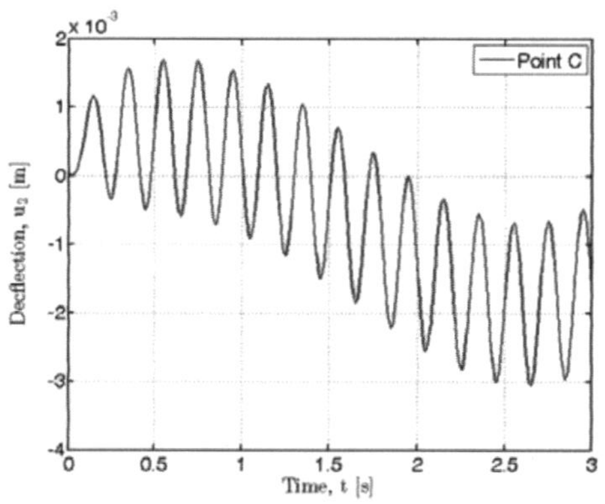

Figure 12. Deflection at point (C).

The deflection at the end point of the structure is more intensive; within the 3 s tolling period the maximum value of deflection at the free end of the structure is about 3 mm, while the maximum deflection at the joint point between the upper and lower columns is about 2 mm.

Case 2: In the next case the rotational spring at the fixed point of the bottom column, point (A), remains the same as before, medium style (H2), while at the joint point between the upper and lower columns, point (B), the rotational spring has hard properties (H3). Loading the system with the same external force and evaluating the simulation under the same time steps, the rotational hysteresis at the bottom point (A) is plotted in figure 13, while the rotational hysteresis of the spring at point (B) is plotted in figure 14.

From the figures it can be seen, that if the joint point between the upper and lower columns is fixed harder than the bottom point of the column, the declination angles at both points (A) and (B) are decreasing.

Journal of Physics: Conference Series **585** (2015) 012005 doi:10.1088/1742-6596/585/1/012005

Considering the deflections of point (B) and at the free end of the column system, point (C), at both places the deflections are decreasing as well, as it can be seen in figure 15 and figure 16. The maximum deflection at point (B) is about 1.2 mm, while at the free end of the column (C) this deflection is about 1.6 mm.

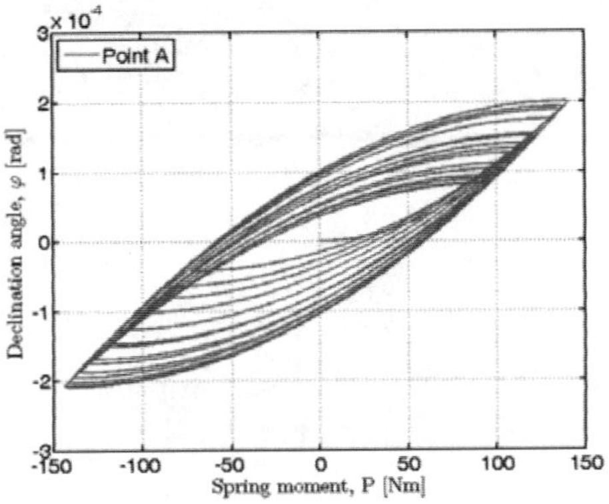

Figure 13. The rotational spring of medium property at point (A).

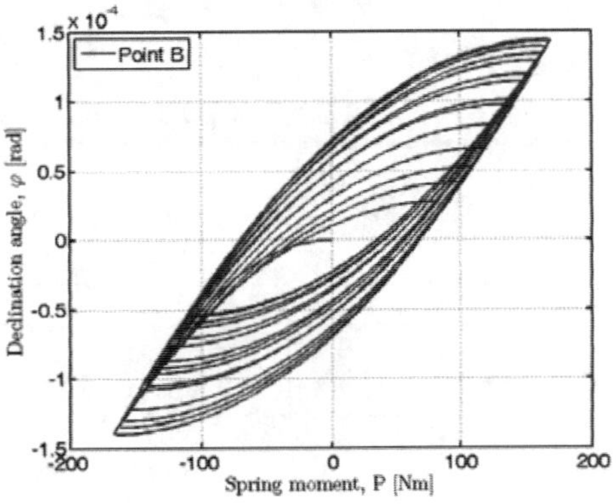

Figure 14. The rotational spring of hard property at point (B).

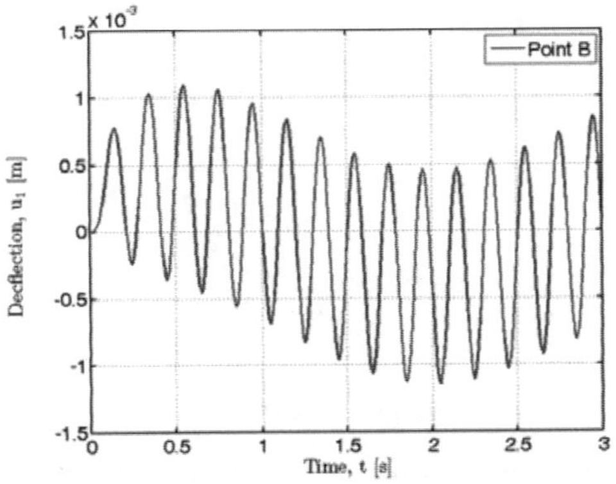

Figure 15. Deflection at the middle point of the column, at point (B).

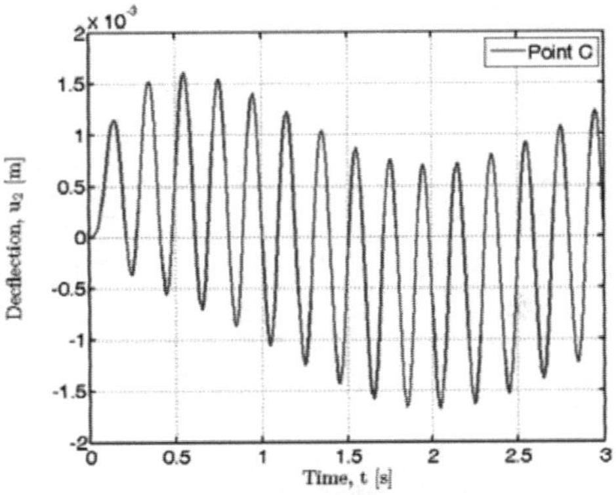

Figure 16. Deflection at the free end of the column, at point (C).

Case 3: Now the rotational spring at the fixed point of the bottom column, point (A), is selected to have weak property (H1), while at the joint point between the upper and lower columns, point (B), the rotational spring has medium property (H2).

For the cycling external force in this case during the motion of the column the hysteresis of the hinges became more stable as it can be seen in figure 17 and figure 18.

Similar stable cycling motion can be observed in the deflections at the joining point between the upper and lower columns, point (B), and at the free end of the column, point (C), as it is plotted in figure 19 and figure 20.

From the figures it can be seen, that however at the fixed point of the bottom column, point (A), the rotational spring has soft property (H1), and at the joint point between the upper and lower columns, point (B), the rotational spring has medium property (H2), the deflections became smaller. At point (B) its maximum value after the first peak is about 1.0 mm, while at point (C) the maximum value of the deflection is about 1.5 mm.

Journal of Physics: Conference Series **585** (2015) 012005 doi:10.1088/1742-6596/585/1/012005

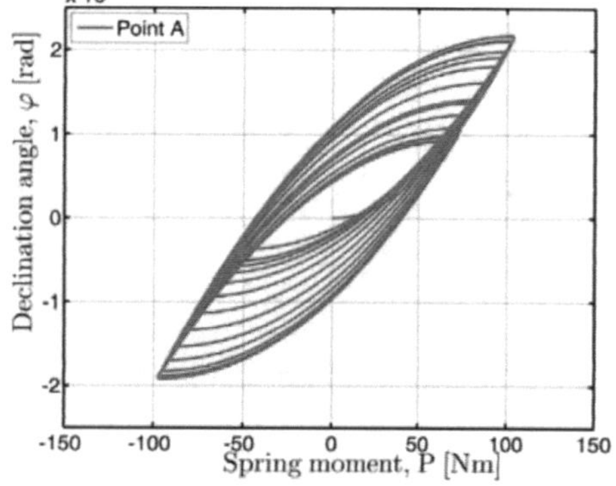

Figure 17. The rotational spring of soft property at point (A).

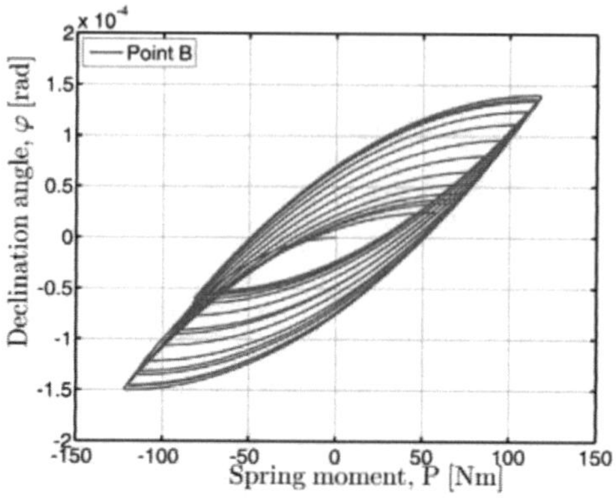

Figure 18. The rotational spring of medium property at point (B).

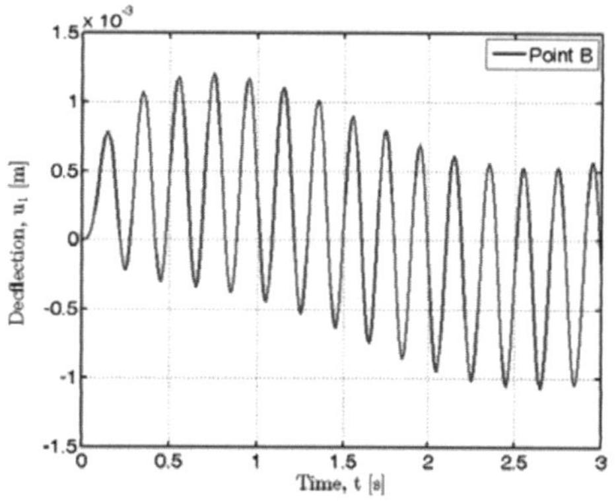

Figure 19. Deflection of the column at point (B).

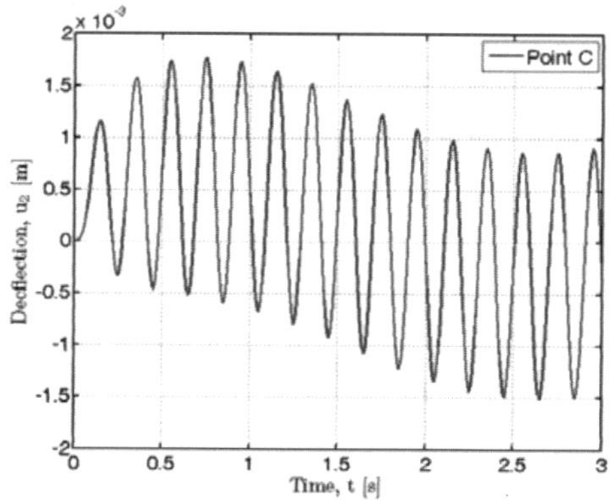

Figure 20. Deflection at the free end of the column, at point (C).

Case 4: The stabilization of the cycling motion can be observed if at the joint point between the upper and lower columns the property of the rotational spring is harder than the property of rotational spring at the fixed point of the bottom column. So, in this case the rotational spring at the fixed point of the lower column, point (A), remain soft (H1), while at the joint point between the upper and lower columns, point (B), the rotational spring is supposed to have hard property (H3).

The hysteresis curves of rotational springs seem to be stabilized during the tolling action as it can be seen at the rotational spring of point (A) in figure 21 and at the rotational spring of point (B) in figure 22.

The hysteresis curves contain some offset during the motion, which can be explained by the mass positioned at the end of each column. The bending moments generated by the gravity force has a constant load on the system and this is added to the cycling load and it results in the offset in the declination and deflection as well.

The deflection at the joint of the columns (point B) and at the free end of the upper column (point C) is plotted in figure 23 and figure 24.

From the figures it can be seen, that under cycling load, at the joining point between the upper and lower columns, point (B), the stabilization of the dynamic system results in the maximum deflection with amplitude about 1.1 mm and at the free point of the column, point (C), the maximum deflection is about 1.6 mm.

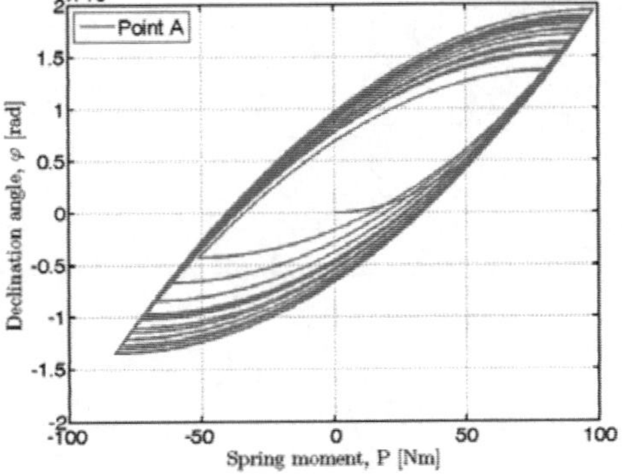

Figure 21. The declination of the soft rotational spring at point (A).

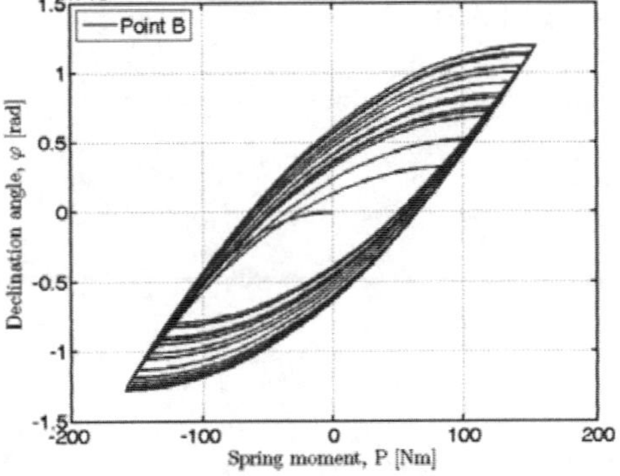

Figure 22. The declination of hard rotational spring at point (B).

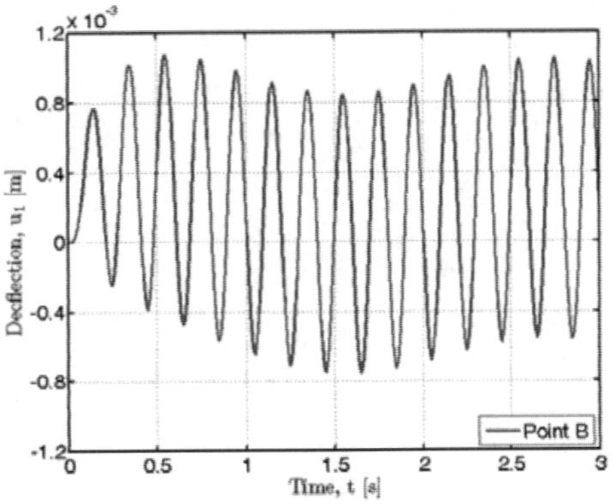

Figure 23. Deflection of the soft rotational spring at the joint point of the cantilevers, in point (B).

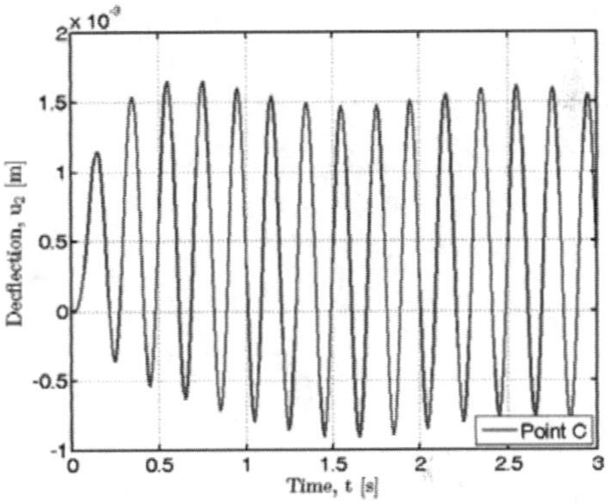

Figure 24. Deflection of the hard rotational spring at he free end of the column system, in point (C).

From the previous studies it can be seen that in all the cases if the rotational spring at the fixed point (A) of the bottom column is even softer than the rotational spring at the joint point between the upper and lower columns, point (B), during the dynamic process the system became stable, see figure 13–14, figure 17-18 and figure 21-22.

The hysteresis curve characterizing the declination of the rotational springs results in stable cycling motion with a small offset. In a similar way, the deflections at the joint of the columns (point B) and at the free end of the upper column (point C) prove to stabilize the elongation as it can be seen in figure 15-16, figure 19-20 and figure 23-24.

Case 5: In this case if the rotational spring at the fixed point of the bottom column (point A) has the same medium property (H2) as the rotational spring at the joining point between the upper and lower columns (point B), some tendency of instability can be observed as it is shown in figure 9-10.

Thus, in the next case the property of the rotational spring at the joining point of the columns (point B) is selected to be softer (H1) than at the fixed point of the bottom column (point A), where the rotational spring is selected to have medium property (H2).

After the numerical simulation at the fixed point of the bottom column (point A) as well as at the joint point of the columns the hysteresis curves definitely exhibit instability as it can be seen in figure 25 and figure 26. According to the soft property of the rotational spring at point B during the cycling load the offset of the hysteresis curves is increasing, finally the failure of the upper column will occur.

Similar instability can be observed in the deflection at the joint of the columns (point B) as well as at the free end of the upper column (point C) as it is plotted in figure 27 and figure 28.

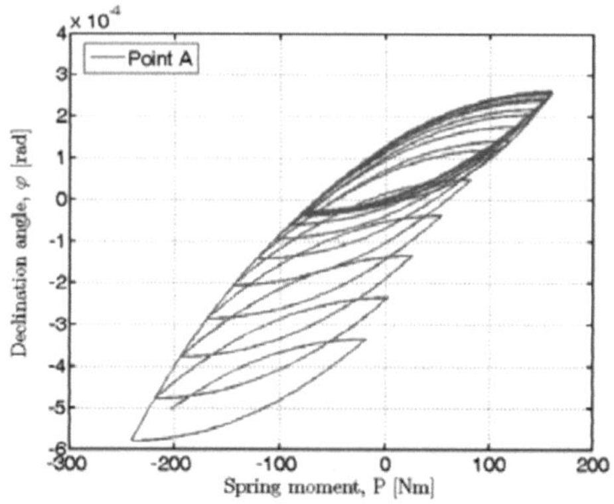

Figure 25. The declination of the medium property of rotational spring (H2) at the fixed point of the column (point A).

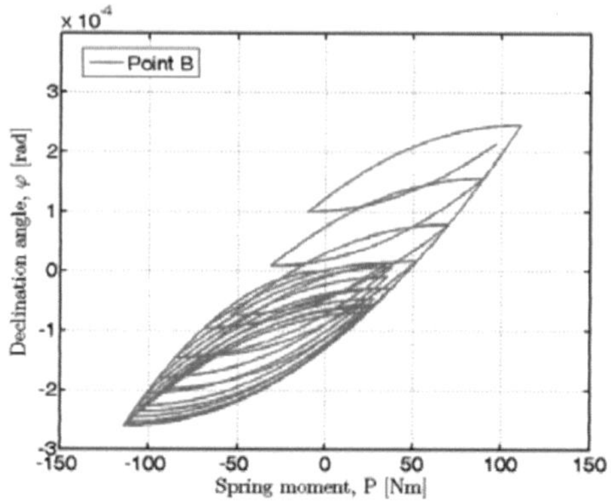

Figure 26. The declination of the soft property of the rotational spring (H1) at the joint point of the columns (point B).

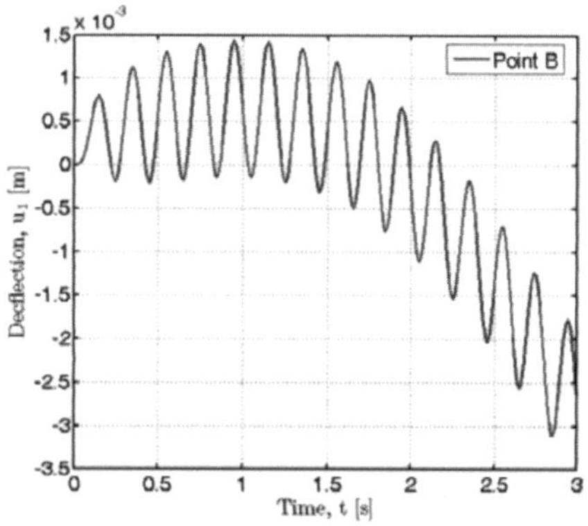

Figure 27. The deflection at the joint of the columns (point B).

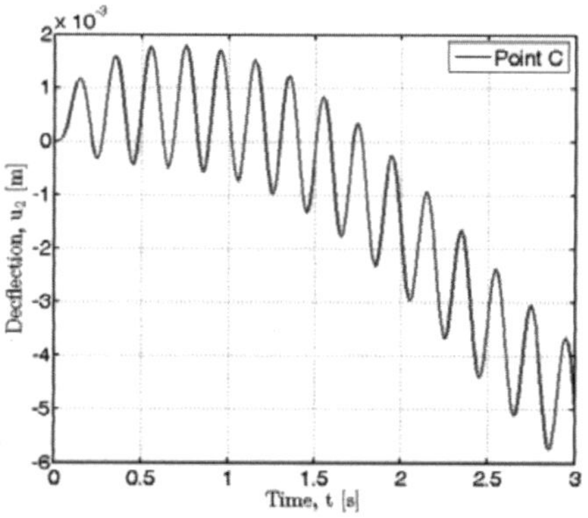

Figure 28. Deflection of the free end of the column (point C).

According to the telescopic construction of the column in this last case instability can be observed in the behaviour of the bell tower. The instability can be observed in the deflection of the whole system as it is shown in figure 29.

7. Conclusion

In this research one column of a telescopic construction of a bell tower has been investigated. The hinges at the fixed support of the bottom column and at the joint of the columns have been modelled with rotational springs of hysteresis characteristics. The mass of the columns and the bell with the fly are concentrated to the top of each column. The tolling process is modelled with a cycling load. The elements of the column are considered to be completely rigid. The non-linear equation of motion is solved by the double application of the Crank-Nicolson schema; the iteration on the non-linear hysteresis characteristic is evaluated by the fix-point technique.

Numerical simulations have been performed assuming the combination of different, soft, medium and hard hysteresis characteristics of hinges. The analytical results prove that in the case when at the fixed point of the bottom column the hinge has softer hysteresis characteristic than at the joint of the columns, the dynamic behaviour of the system can be stable. However, in the case when the hinge at the joint of the columns is softer than the one at the fixed support of the bottom column, the instability of the system can be observed.

Figure 29. The instability property of the construction under cycling load.

8. References

[1] Stoner EC and Wohlfarth EP 1991 Reprinted in *IEEE Tr. on Magn.* **27** 3457
[2] Jiles DC and Atherton DL 1986 *J. of Magn. and Mag. Materials* **61** 48
[3] Chikazumi S and Charap SH 1964 *Physics of Magnetism* (J. Wiley)
[4] Krasnosel'skii MA and Pokrovskii AV 1983 *System with Hysteresis* (in Russian) (Moscow: Nauka)
[5] Mayergoyz ID 1991 *Mathematical Model of Hysteresis* (Springer)
[6] Visintin A 1994 *Differential Models of Hysteresis* (Springer)
[7] Brokatte M and Sprekels J 1996 *Hysteresis and Phase Transitions* (Springer)
[8] Ivanyi A 1997 *Hysteresis Models in Electromagnetic Computation* (Budapest: Akademiai Kiado)
[9] Bertotti G 1998 *Hysteresis in Magnetism, for Physicists, Material Scientists, and Engineers* (NY: Academic Press)
[10] Della Torre E 1999 *Magnetic Hysteresis* (NY: IEEE Press)
[11] Díaz C, Martí P, Victoria M and Querin OM 2011 *J. of Constructional and Steel Research* **67** 741
[12] Kwofie S 2003 *Material and Engineering* **A357** 86
[13] RambergW and Osgood WR 1943 Technical Note No. 902 *National Advisory Committee for Aeronautics* (Washington DC)
[14] Demartino A, Landolfo R and Mazzolani FM 1990 *Materials and Structures* **23** 59
[15] Skelton R P, Maier HJ and Christ HJ 1997 *Material Science in Engineering* **A238** 377
[16] Mostaghel N and Byrd RA 2002 *Int. J. of Non-Linear Mechanics* **37** 1319
[17] Niesłony A, el Dsoki C, Kaufmann H and Krug P 2008 *Int. J. of Fatigue* **30** 1967
[18] Ni YQ, Wang JY and Ko JM 1999 Advanced method for modelling hysteretic behaviour of semi-rigid joints *Proceedings of the Second Int. Conf. on Advances in Steel Structures* (Hong Kong, China, 15-17 December 1999) (in *Advances in Steel Structures*) ed SL Chan and JG Teng (Elsevier) vol 1 pp 331–338
[19] Richard MR and Abbott BJ 1975 *J. Eng. Mech. Div. ASCE* 1975 **101** 4 511
[20] Smith RC, Seelecke S, Dapino M and Ounaies Z 2006 *J. of the Mechanics and Physics in Solids* **54** 46
[21] Anglada-RiveraJ, Padovese LR and Capó-Sánchez J 2001 *J. of Magn. and Mag. Materials* **231** 299

[22] Steven KJ 2000 *NDT&E International* **33** 111
[23] Permiakov V, Dupré L, Pulnikov A and Melkebeek J 2004 *J. of Magn. and Mag. Materials* **272–276** e2553
[24] Aleshin V and Van Den Abeele K 2005 *J. of the Mechanics and Physics of Solids* **5553** 795
[25] Bolshakov GV and Lapokov AJ 1996 *J. of Magn. and Mag. Materials* **162** 112
[26] BerquistA and Engdahl G 1996 *IEEE Trans. on Magn.* **27** 112
[27] BachmannB and Bachman Z 2010 *Pollack Periodica* **5** 3 19
[28] Kuczmann M and Ivanyi A 2008 *The Finite Element Method in Magnetics* (Budapest: Akademiai Kiado)
[29] Chan SL and Chui PPT 2000 *Nonlinear Static and Cycling Analysis of Steel Frames with Semi-rigid Connections* (Elsevier)
[30] Maroti Gy 2011*Pollack Periodica* **6** 1 141
[31] Krejci P 1988 *Application of Mathematics* **33** 145
[32] Ivanyi A 2003 *Continuous and Discrete Simulations in Electrodynamics* (Budapest: Akademiai Kiado)
[33] Bottachio O, Chiampi M and Ragusa C 2003 *IEEE Trans. on Magn.* **34** 1179
[34] KuczmannM 2011 *Physica B* Condensed Matter **406** 1403

Stabilization of a system with saturating, non-monotone hysteresis and frequency dependent power losses by a PD controller

D B Ekanayake[1] and R V Iyer[2]

[1] Mathematics Department, Western Illinois University, Macomb, IL 61455, USA. [2] Dept. of Mathematics and Statistics, Texas Tech University, Lubbock, TX 79409, USA.

E-mail: DB-Ekanayake@wiu.edu, ram.iyer@ttu.edu

Abstract. We prove the closed loop stability of a PD controller for certain systems with saturating, non-monotone hysteresis and frequency dependent power losses. Most controllers use inverse compensators to cancel out actuator hysteresis nonlinearity. We show that we can achieve stability of the closed-loop system without an explicit inverse computation (using least squares minimization or otherwise).

1. Introduction

Magnetic and smart actuators are smaller and more energy efficient than conventional actuators, and are useful in number of precision control applications [1, 2, 3, 4, 5]. However, they are difficult to control to achieve a given objective due to the occurrence of hysteresis loops, memory effects, and saturation, as well as rate dependent power losses. Motivated by such hysteretic systems, in this paper, we discuss how to develop a proportional derivative (PD) control for saturating, non–monotone hysteretic systems with nonlinear power losses.

The controller strategy discussed in this paper is derived for tracking control, that is, for forcing the system output to follow a specific trajectory. However, we only discuss the stabilization problem in this paper. The system has two measured feedbacks that can be used for control: the error output, which we want to minimize, and another measured output. In the case of magnetic and smart actuators, the other output is the induced voltage (electro-motive force) of the actuator windings. Induced voltage of the actuator windings can be obtained from the input current and voltage (shown in [6, 7]), and the error output is derived from the actuator tip displacement. The use of two feedbacks makes the control design entirely different from those used in the literature for control of hysteretic systems.

Tracking control of hysteretic systems has received wide attention in recent years. Tracking control methods presented in the literature include inverse compensation of hysteresis [9, 10, 11], adaptive control [12, 13, 14], integral control [15, 16, 17], passivity–based control [18, 19], monotonicity-based control [20, 21], and hybrid control, most of which are model–dependent [22, 23].

The most common control method discussed in literature is inverse hysteresis compensation, where an approximate inverse of the hysteresis is employed to suppress the hysteresis nonlinearity in the system [8, 9, 11, 24, 25]. These controllers are model–based and applicable for hysteretic

Journal of Physics: Conference Series **585** (2015) 012006 doi:10.1088/1742-6596/585/1/012006

systems at low frequencies [9]. Another widely investigated strategy is integral, proportional, and derivative (PID) control [15, 17], derived from the output error. These controllers have received much attention because of the simplicity and availability of PID modules. A key condition that aids in the development of these controllers is that hysteresis is piecewise monotonic. In literature, these controllers are employed for systems in which hysteresis is the only nonlinearity, or the PID controllers are combined with model–based controllers [22, 23]. Adaptive control is another common method for controlling hysteretic systems. Some adaptive controllers are based on inverse hysteresis compensation [12, 26], while others are robust adaptive control schemes without the hysteresis inverse [14, 27]. They are primarily utilized for non–saturating hysteresis operators without minor–loop closure behavior.

1.1. Novelty of the proposed stabilizing controller

None of the aforementioned control schemes are suitable for non-monotone hysteretic systems showing minor-loop closure and saturation. Magnetic and magnetostrictive actuators exhibit such complex hysteresis with minor-loop closure, as well as other nonlinearities such as eddy current and residual losses, which increase with the operating frequency. These actuators become saturated and some actuators relate the output of the hysteresis to the system output via a square map, such that the system is not monotone. These behaviors limit the usefulness of conventional control schemes. Due to the limitations of both the models and the control schemes, tracking control is limited to low frequency ranges (less than 200 Hz [9]) and low amplitude signals, which prevents the actuators from becoming saturated. Hence, to control these systems in their full frequency and amplitude ranges, a new scheme is required to achieve tracking control.

Existing control schemes for hysteretic systems use only the error output, to be minimized, to derive the control signal. For example, existing control schemes for magnetostrictive actuators use only the actuator tip displacement measurements as input for the control scheme [24]. However, when a current is applied to a magnetostrictive actuator, measurements can be taken for the voltage induced by the actuator, as well as for the displacement of the tip. To achieve precision control of these complex systems, it is important to include all measurements that can easily be obtained and that reflect properties of the hysteretic system. Hence, we use two feedbacks for control. In the following analysis, we show that, given any $\varepsilon > 0$, the output $y(t)$ is ultimately bounded by ε (that is, $\limsup_{t\to\infty} |y(t)| \le \varepsilon$), in the absence of disturbance.

2. System with Hysteresis

In this paper, we consider hysteretic systems modelled for almost every $t \in [0,\infty)$ by nonlinear functional differential equations of the form [28]

$$\dot{y}(t) + \alpha y(t) = aM^2(t) + bM(t) + \Delta(t) \quad (1)$$

$$\frac{d}{dt}[u + M](t) = \gamma x(t) \quad (2)$$

$$\theta(x)(t) + \beta u(t) = \widehat{u}(t) \quad (3)$$

$$M(t) = \Gamma[u\ ;\ \psi_{-1}](t) \quad (4)$$

$$y(0) = y_0, \qquad u(0) = u_0, \quad (5)$$

where $\widehat{u}(t)$ is the system input, $x(t)$ is an output and $y(t)$ is the output to be regulated. Functions $u(t)$ and $M(t)$ represent two internal states. The system of equations (1-5) may be used to model a magnetostrictive actuator with inertia ignored (set $m = 0$ in the model of [28]). In this case, $x(t)$ corresponds to the induced voltage, M represents the average magnetization along the axis, y corresponds to the displacement of the actuator tip, u corresponds to the axial average magnetic field, and $\widehat{u}$ corresponds to the applied voltage. Equation (2) is equivalent to Faraday's Law. Operator $\Gamma[\cdot\ ;\ \psi_{-1}]$ is the hysteresis operator with initial memory state ψ_{-1}. The function

Δ represents a actuator disturbance signal, while θ is a nonlinear function which represents eddy current and excess losses for magnetic systems.

We assume hysteresis can be represented by an operator of Preisach type. In the literature, Presach operator has been shown to be a well–suited approximation for magnetic and smart actuators [29, 30, 31]. The following definition of Preisach operator may be found in Brokate and Sprekels [32]. Suppose Ψ_0 is the space of admissible memory curves:

$$\Psi_0 := \{\psi \mid \psi : \mathbb{R}_+ \to \mathbb{R},\ |\psi(r) - \psi(\bar{r})| \le |r - \bar{r}|\ \forall r, \bar{r} \ge 0, R_{supp}(\psi) < \infty\},$$

where $R_{supp}(\psi) := \sup\{r \mid r \ge 0, \psi(r) \ne 0\}$. The Preisach operator $\Gamma[u\,;\psi_{-1}](t)$, with the initial memory curve ψ_{-1}, is defined with an output map $Q : \Psi_0 \to \mathbb{R}$ of the form:

$$Q(\psi) = \int_0^\infty q(r, \psi(r))dr + w_{00}, \quad \text{where} \quad q(r, s) = 2\int_0^s \omega(r, \sigma)d\sigma, \tag{6}$$

with $w_{00} = \int_0^\infty \int_{-\infty}^0 \omega(r, s)dsdr + \int_0^\infty \int_0^\infty \omega(r, s)dsdr$. Here, $\omega \in L^1_{loc}(\mathbb{R}_+ \times \mathbb{R})$ is the Preisach density function. For piecewise monotone functions u, the Preisach operator is defined to be: $\Gamma[u; \psi_{-1}(t) \triangleq Q(\psi(t)(r))$, where $\psi(t)(r) = \mathcal{F}_r[u(t); \psi_{-1}(r)]$ and $\mathcal{F}_r$ is the Play operator with parameter r [32].

3. Statement of the problem

The objective is to design a controller such that all the internal signals u, x, y of the system (1-5) are bounded for all $t \ge 0$ even in the presence of the disturbance Δ. Furthermore, all signals must be uniquely specified. Finally, when the disturbance $\Delta = 0$, for arbitrary $\varepsilon > 0$, the controller must have the property that $|y(t)|$ is ultimately bounded by ε (that is, $\limsup_{t\to\infty} |y(t)| < \varepsilon$) by proper choice of parameters. We introduce the following assumptions.

$\mathcal{H}1$: The Preisach operator $\Gamma[\cdot\,; \psi_{-1}]$ is output-rate dissipative (counter-clockwise dissipative), piecewise monotone, and Lipschitz continuous on $C[0, T]$ and $W^{1,1}(0, T)$ where $T > 0$ is any positive real number.

$\mathcal{H}2$: There exist real numbers $\Gamma_{sat1}, \Gamma_{sat2} > 0$ such that $-\Gamma_{sat1} = \inf_{\psi\in\Psi_0} Q(\psi)$ and $\Gamma_{sat2} = \sup_{\psi\in\Psi_0} Q(\psi)$.

$\mathcal{H}3$: $\theta(\cdot)$ is continuous and strictly monotone increasing; $\theta(0) = 0$.

$\mathcal{H}4$: $a \ge 0$, while b, α, β, and γ are positive. The disturbance signal satisfies $\Delta \in L^\infty[0, \infty)$. Furthermore, $\|\Delta\|_\infty \le \eta$, and $\eta < \frac{b^2}{4a}$. The initial value parameter y_0 lies in a known bounded set.

Hypothesis $\mathcal{H}1$ is a typical condition which guarantees the thermodynamic consistency of hysteresis operators. In particular, it reflects the fundamental energy dissipation properties of hysteresis. Hypotheses $\mathcal{H}2$ states that the hysteresis operators in this study are allowed to saturate. The next hypothesis, $\mathcal{H}3$, describes the behavior of function θ. The monotonicity represents the increase of power losses as the magnitude of x increases. This hypothesis is the result of modeling frequency dependent excess losses in magnetic systems [6, 28].

4. Feedback Stabilization

The following result is an extension of Corollary 2.4.12 in Brokate and Sprekels [32]. It establishes the Lipschitz continuity of the operator $\Gamma \circ (I + \Gamma)^{-1} : W^{1,1}(0, T) \to W^{1,1}(0, T)$. It is useful to prove the existence and uniqueness of solutions to (1-5). If I is an interval in $\mathbb{R}$, $C_I[0, T] = \{u \in C[0, T] \mid u(t) \in I;\ t \in [0, T]\}$. We assume that for $\psi_{-1} \in \Psi_0$, if $u \in C_I[0, T]$, then $\Gamma[u; \psi_{-1}] \in C_J[0, T]$.

Lemma 4.1 *Suppose $\Gamma[u;\psi_1]$ is a Preisach operator generated by a nonnegative bounded density function $\omega(r,s)$ with compact support, where $\psi_{-1}\in\Psi_0$. Furthermore, suppose that $\frac{\partial}{\partial s}\omega(r,s)$ is measurable and bounded on bounded subsets of $\mathbb{R}_+\times\mathbb{R}$. Then, for any $T>0$, the following assertions hold:*

(a) $\Gamma[\cdot;\psi_{-1}]$ is Lipschitz continuous on $C[0,T]$ and on bounded subsets of $W^{1,1}(0,T)$.

(b) $(\mathcal{I}+\Gamma)[\cdot;\psi_{-1}]: C_I[0,T]\to C_J[0,T]$ is invertible. $(\mathcal{I}+\Gamma)^{-1}[\cdot;\psi_{-1}]$ is Lipschitz continuous on $C_J[0,T]$ and on bounded subsets of $W^{1,1}(0,T)$.

(c) $\Gamma\circ(\mathcal{I}+\Gamma)^{-1}[\cdot;\psi_{-1}]$ is Lipschitz continuous on $C_J[0,T]$ and on bounded subsets of $W^{1,1}(0,T)$.

The proof of the above result may be shown using the results proved in [32].

The stabilizing control signal can be derived either using a PI controller or a PD controller. In this paper, we restrict our attention only to the PD controller configuration. Control input $\widehat{u}(t)$ is derived from signals y and x:

$$\widehat{u}(t) = -\widehat{k}_P x - k_P y - k_D\dot{y}. \tag{7}$$

Next we obtain the closed loop system as follows. Equation (2) can be expressed in integral form as

$$u + \Gamma[u\,;\psi_{-1}] = f(x), \tag{8}$$

where, $f(x) := \gamma\int_0^t x d\tau + u_0 + \Gamma_0$. By the invertibility of $\mathcal{I}+\Gamma$, $u = (\mathcal{I}+\Gamma)^{-1}f(x)$. Hence $\Gamma[u\,;\psi_{-1}]$ can be rewritten in terms of x as $W[f(x)\,;\psi_{-1}] := \Gamma\left[(\mathcal{I}+\Gamma)^{-1}f(x)\,;\psi_{-1}\right]$. From Equation (8),

$$u = \gamma\int_0^t x d\tau - W[f(x)\,;\psi_{-1}] + u_0 + \Gamma_0. \tag{9}$$

Then, Equation (9) combined with system of equations (1-5) yields the closed loop system:

$$\begin{aligned} \dot{y} + \alpha y &= g\left(W[f(x)\,;\psi_{-1}]\right) + \Delta(t) && (10)\\ \varphi(x(t)) + \beta\gamma\int_0^t x(\tau)d\tau &= (k_D\alpha - k_P)y(t) + G(t_0;x)(t), && (11)\end{aligned}$$

where

$$\begin{aligned} \varphi(x) &= \widehat{k}_P x + \theta(x),\\ g(\cdot) &= a(\cdot)^2 + b(\cdot), \qquad \text{and}\\ G(t_0;x)(t) &= \beta W[f(x)\,;\psi_{-1}](t) - k_D g\left(W[f(x)\,;\psi_{-1}](t)\right) - \beta(u_0+\Gamma_0) - k_D\,\Delta(t).\end{aligned}$$

Due the semi-group property satisfied by Γ, we also have for any $0\le t_1\le t$,

$$\begin{aligned} &\varphi(x) + \beta\gamma\int_{t_1}^t x(\tau)d\tau = (k_D\alpha - k_P)y + G(t_1;x)(t),\\ &G(t_1;x)(t) = \beta W[f(x);\psi_{-1}](t) - k_D g\left(W[f(x);\psi_{-1}]\right) - \beta(u(t_1) + \Gamma[u;\psi_{-1}](t_1)) - k_D\,\Delta(t).\end{aligned}$$

This observation is useful in Section 4.1.

Next we need to show that the above system possesses a solution such that $y\in W^{1,\infty}[0,\infty)$ satisfies the differential equation almost everywhere. Further, all the states must remain bounded for all $t\in[0,\infty)$. The next section establishes existence, uniqueness and stability of a solution to Equations (10)-(11). We then show that for the disturbance-free case $\Delta=0$, given any $\varepsilon>0$, we can make $|y(t)|$ ultimately bounded by ε by careful choice of parameters $\widehat{k}_P, k_P, k_D$.

4.1. Existence and uniqueness results

First, we determine the specific properties required of function $\varphi(x)$. Proof of the following lemma is a direct consequence of the inverse function theorem.

Lemma 4.2 *For each $x \in \mathbb{R}$, the function $\varphi := \widehat{k}_P x + \theta(x)$ is strictly monotone. Further, φ^{-1} is well-defined, strictly monotone, and Lipschitz continuous with Lipschitz constant $\frac{1}{\widehat{k}_P}$.*

This lemma shows that by selecting $\widehat{k}_P > 0$, we obtain the Lipschitz continuity of φ, which is important to prove existence and uniqueness of the closed-loop system.

Since function $g(M(t)) = aM^2(t) + bM(t)$ is non-monotone (Figure 1), two values of $M(t)$ achieve the same value of $g(M(t))$ and, hence, the same output $y(t)$. However, to avoid ambiguity in the control design, the value of $M(t)$ must be restricted to an interval such that a unique value of M corresponds to each value of $g(M)$. If $b = 0$, then $g(M)$ can only take on non-negative values and it can quickly be seen that (1) cannot be stabilized, that is $\limsup_{t\to\infty} |y(t)|$ cannot be made to be less than any given $\varepsilon > 0$. If $a = 0$, then the graph of $g(M)$ is monotone on $[-\Gamma_{sat1}, \Gamma_{sat2}]$. If $a > 0$ and $b > 0$, then the initial value of M, that is, Γ_0 must be away from the turning point $-\frac{b}{2a}$ in order that we have local uniqueness.

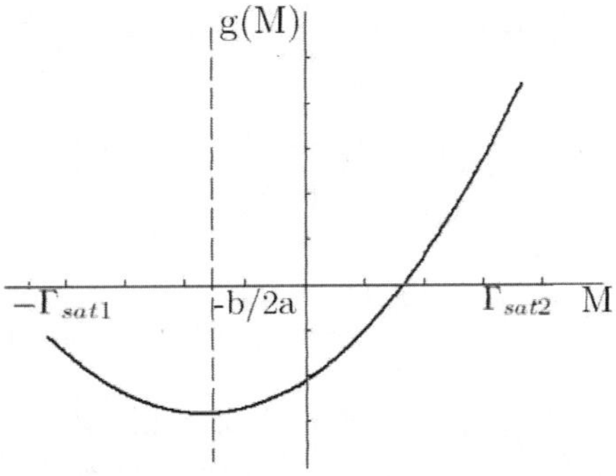

Figure 1. Function $g(M)$

Theorem 4.1 (Local Existence and Uniqueness) *Suppose hypotheses $\mathcal{H}1$ - $\mathcal{H}4$ hold for all $T > 0$. Further suppose that $\Gamma_0 \in (\max\{-\frac{b}{2a}, -\Gamma_{sat1}\}, \Gamma_{sat2}]$. There exists $\tau > 0$ such that the system (1) - (5) and (7) has a unique solution $(B, y) \in W^{1,\infty}(0,\tau) \times W^{1,\infty}(0,\tau)$, where $B = u + M$.*

Proof With $B = u + M$, $\varphi(x) = \widehat{k}_P x + \theta(x)$, $V = (I + \Gamma)^{-1}[\cdot; \psi_{-1}]$ and $g(x) = ax^2 + bx$, we have $u = V(B)$. The combined system (1) - (5) and (7) is:

$$\begin{aligned} \dot{B} &= \gamma\varphi^{-1}(-\beta V(B) - (k_P - k_D\alpha)y - k_D\, g((\Gamma \circ V)(B)) - k_D\,\Delta(t)) \\ \dot{y} &= -\alpha y + g((\Gamma \circ V)(B)) + \Delta(t). \end{aligned}$$

The above equations are in the form $\dot{w} = F(t, w, V[\mathcal{S}(w)], W[\mathcal{S}(w)])$ where $w = (B, y) \in \mathbb{R}^2$, $\mathcal{S}[w] = B$ is a projection operator, and $W = \Gamma \circ V$. Note that F is Lipschitz continuous as function of the last three arguments, and measurable as a function of t. As V and $\Gamma \circ V$ are Lipschitz continuous from $W^{1,1}(0,T)$ to $W^{1,1}(0,T)$ by Lemma 4.1, the claim follows from the same argument as in Theorem 3.1.1 in [32]. □

Theorem 4.1 concludes the existence of a local solution on a sufficiently small interval $[0, \tau)$. Our goal is to show existence and uniqueness on $[0, \infty)$. However, extension of the solution presents a problem because $M(t)$ might pass through the turning point $-\frac{b}{2a}$ after which multiple solutions might exist. The following two lemmas together show that this cannot happen by proper choice of parameters k_P and k_D.

Lemma 4.3 *Suppose that the conditions of Theorem 4.1 hold. Let a unique solution to (10) - (11) exist on the interval* $[0, \tau)$*. Then, there exists bounds independent of parameters* k_P *and* k_D *for both* y *and* x *on* $[0, \tau)$*.*

Proof By Theorem 4.1, there exists a solution on a sufficiently small interval $[0, \tau)$. From Equation (10), for $t \in [0, \tau)$:

$$\begin{aligned} \dot{y}(t) + \alpha y(t) &= g(W[f(x)])(t) + \Delta(t) \\ \implies \quad y(t) &= e^{-\alpha t}\left[\int_0^t e^{\alpha\tau} g(W[f(x)])(\tau)d\tau + \int_0^t e^{\alpha\tau}\Delta(\tau)d\tau + y_0\right]. \end{aligned}$$

Since W is saturating (by Hypothesis $\mathcal{H}2$) and Δ is essentially bounded on $[0, \infty)$ (by Hypothesis $\mathcal{H}4$),

$$|y(t)| \le e^{-\alpha t}\int_0^t N e^{\alpha\tau} d\tau + e^{-\alpha t}|y_0|,$$

for some constant N. Hence,

$$\forall\, t \in [0, \tau), \qquad |y(t)| \le N/\alpha + |y_0|. \tag{12}$$

With $z = \varphi(x)$, (11) becomes:

$$z(t) \;=\; G\left(0; \varphi^{-1}(z)\right)(t) - \beta\gamma \int_0^t \varphi^{-1}(z(\tau))d\tau + (k_D\alpha - k_p)y(t), \quad t \in [0, \tau). \tag{13}$$

Due to Hypotheses $\mathcal{H}2$ and $\mathcal{H}4$, there exists $L > 0$ such that $\forall\, t \in [0, \tau)$, $|G(0; \varphi^{-1}(z))(t)| < L$. Since y is bounded (by Claim 1), there exists M such that $|(k_D\alpha - k_P)y(t)| < M$ for all $t \in [0, \tau)$. By (11), $z(0) < L + M$.

We claim that $|z(t)| < L + M$ for all $t \in [0, \tau)$. Suppose that this is not the case. Then there exists $t^* \in [0, \tau)$ such that $|z(t^*)| \ge L + M$. Consider the case $z(t^*) \ge L + M$. Suppose $z(t) > 0$ for all $t \in (t^{**}, t^*]$, where $z(t^{**}) = 0$. Since $\varphi^{-1}(0) = 0$ and φ^{-1} is strictly monotone increasing, $x(t) = \varphi^{-1}(z)(t) > 0$ for $t \in (t^{**}, t^*]$. By the semi-group property of Γ (see the comment following (11)),

$$\begin{aligned} L + M \le z(t^*) &= G\left(t^{**}; \varphi^{-1}(z)\right)(t^*) - \beta\gamma\int_{t^{**}}^{t^*} \varphi^{-1}(z(\tau))d\tau + (k_D\alpha - k_P)y(t^*) \\ &< L + M, \end{aligned}$$

which is a contradiction. By a similar argument, we can obtain $z(t) > -L - M$, concluding the proof of the claim. Hence,

$$\forall\, t \in [0, \tau), \qquad |x(t)| \le \varphi^{-1}(L + M).\square \tag{14}$$

For the case, $a > 0$ and $b > 0$, we may assume $b/2a < \Gamma_{sat1}$, since otherwise $[-\Gamma_{sat1}, \Gamma_{sat2}] \subset [-b/2a, \Gamma_{sat2}]$ and $M(t)$ is uniquely determined as desired.

Denote $\pi = [-b/2a, \Gamma_{sat2}]$. Lemma 4.4 establishes that appropriate choices for the gain parameters force the controller to maintain $M(t) \in [-b/2a, \Gamma_{sat2}]$.

Lemma 4.4 *Suppose that the conditions of Theorem 4.1 hold. Let a unique solution to (10) - (11) exist on the maximal interval* $[0,\tau)$. *Then* $\forall\ k_0 > 0\ \exists\ k_{P0} > 0 \ni$ *if* k_P, k_D *satisfy* $k_P \geq k_{P0}$ *and* $k_D \in [\frac{k_P}{\alpha}, k_0 + \frac{k_P}{\alpha}]$, *then* $M(t) \in \pi$ *for all* $t \in [0,\tau]$. *Furthermore, if* $M(\tau) = -\frac{b}{2a}$ *then* $\dot{M}(\tau) \geq 0$.

Proof By the conditions of Theorem 4.1, $M(0) \in \text{Interior}(\pi)$. Given $k_0 > 0$, pick $k_{P0} = \frac{\alpha(\beta d_1 + k_0 d_2)}{\frac{b^2}{4a} - \eta}$. Suppose k_D satisfies $0 \leq k_D - \frac{k_P}{\alpha} \leq k_0$.

As $[0,\tau)$ is the maximal interval of existence and uniqueness, we have either a loss of existence (solution becomes unbounded) or loss of uniqueness as $t \to \tau$. Lemma 4.3 shows that all signals exist at $t = \tau$. Hence, the only possibility is that there is a loss of uniqueness for $t > \tau$. This may happen if $M(\tau) = -\frac{b}{2a}$. If $\dot{M}(\tau)$ does not exist, then M is increasing at τ. Suppose $\dot{M}(\tau)$ exists. We only need to show that we can find parameters k_P and k_D such that $\dot{M}(\tau) \geq 0$.

From (3) and (7),

$$\beta\, u(t) = -\widehat{k}_P x(t) - \theta(x)(t) - k_P y(t) - k_D \dot{y}(t).$$

It was shown in the proof of Lemma 4.3 that $x(\tau)$, $y(\tau)$ are bounded and the bounds are independent of parameters $\widehat{k}_P$, k_P and k_D. It follows that $\forall t \in [0,\tau]$, $|u(t)| \leq d_1$ for some constant d_1 that is independent of gain parameters.

Now, $\forall t \in [0,\tau]$, $M(t) \in \pi$, which implies that $|\dot{y}(t)| < d_2$ for some constant d_2 that is independent of parameters $\widehat{k}_P$, k_P and k_D. As $M(\tau) = -b/2a$,

$$\dot{y}(\tau) + \alpha y(\tau) = -b^2/4a + \Delta(\tau).$$

Denote as before, $\varphi(x)(\tau) = \widehat{k}_P x(\tau) + \theta(x)(\tau)$.

We rewrite (3) and (7) at time τ by adding and subtracting $-\frac{k_P}{\alpha}\,\dot{y}(\tau)$ to the right hand side, and collecting similar terms:

$$\begin{aligned}
\varphi(x)(\tau) &= -\beta\, u(\tau) - \frac{k_P}{\alpha}\,\dot{y}(\tau) - k_P y(\tau) - k_D \dot{y}(\tau) + \frac{k_P}{\alpha}\,\dot{y}(\tau) \\
&= -\beta\, u(\tau) - \frac{k_P}{\alpha}\,(\dot{y}(\tau) + \alpha y(\tau)) + \left(-k_D + \frac{k_P}{\alpha}\right)\dot{y}(\tau) \qquad (15)\\
&\geq -\beta\, d_1 + \frac{k_P}{\alpha}\left(\frac{b^2}{4a} - \eta\right) - \left(k_D - \frac{k_P}{\alpha}\right) d_2. \\
&\geq -\beta\, d_1 + \frac{k_P}{\alpha}\left(\frac{b^2}{4a} - \eta\right) - k_0\, d_2 \qquad (16)\\
&= \frac{1}{\alpha}\left(\frac{b^2}{4a} - \eta\right)(k_P - k_{P0}) \quad \geq 0 \qquad (17)
\end{aligned}$$

By Hypothesis $\mathcal{H}3$, $x(\tau) \geq 0$. By Hypothesis $\mathcal{H}1$ and Equation (2), $\dot{M}(\tau) \geq 0$. □

Having shown that M is uniquely determined even if it reaches the boundary of π, we can now prove the following theorem.

Theorem 4.2 (Global Existence and Uniqueness) *Suppose hypotheses* $\mathcal{H}1$-$\mathcal{H}4$ *hold for all* $T > 0$ *and* $\Gamma_0 \in (\max\{\frac{-b}{2a}, -\Gamma_{sat1}\}, \Gamma_{sat2}]$. *Suppose the parameters* k_0, k_P *and* k_D *satisfy the conditions specified in Lemma 4.4. Then, for any* $T > 0$, *(10), and (11) have a unique solution* $(B, y) \in W^{1,\infty}(0,T) \times W^{1,\infty}(0,T)$, *where* $B = u + M$. *Furthermore,* $x \in BC[0,\infty)$, $y \in W^{1,\infty}(0,\infty)$ *and* $\|y\|_\infty + \|\dot{y}\|_\infty \leq K$ *for some* $K < \infty$ *independent of parameters* k_P, k_D, *and* $\widehat{k}_P$.

Journal of Physics: Conference Series **585** (2015) 012006 doi:10.1088/1742-6596/585/1/012006

Proof Suppose $[0, \tau)$ is the maximal interval of existence and uniqueness of the solution to (10) and (11), where $\tau < \infty$. We will prove a contradiction.

By Lemma 4.3 $y(t)$ and $x(t)$ are uniformly bounded on $[0, \tau)$.

Therefore, (8) yields,

$$\forall\, t \in [0, \tau), \qquad |B(t)| \leq \tau\, \gamma\, \varphi^{-1}(L + M) + |u_0 + \Gamma_0|. \tag{18}$$

Using Theorem 4.1, we see that the solution may be extended to an interval larger than $[0, \tau)$, and the extension is unique by Lemma 4.4. This contradicts the maximality of that interval $[0, \tau)$. Hence, the solution exists on the half-line $[0, \infty)$. We also see that for any $T > 0$, $(B, y) \in W^{1,\infty}(0, T) \times W^{1,\infty}(0, T)$.

The claim that $x \in BC[0, \infty)$ follows from the proof of Lemma 4.3. The claim that $y \in W^{1,\infty}[0, \infty)$ follows immediately from Equation (10) by observing that,

$$\|\dot{y}\|_\infty \leq N + \alpha|y_0| + a \max\{\Gamma_{sat1}^2, \Gamma_{sat2}^2\} + b \max\{\Gamma_{sat1}, \Gamma_{sat2}\} + \eta.$$

Select

$$K = (\frac{1}{\alpha} + 1)\,(N + \alpha|y_0|) + a \max\{\Gamma_{sat1}^2, \Gamma_{sat2}^2\} + b \max\{\Gamma_{sat1}, \Gamma_{sat2}\} + \eta\}.$$

□

4.2. Proof of Stabilization

The next objective is to show that the control signal given by Equation (7) achieves stabilization. In the previous subsection, we saw that by selecting $\widehat{k}_P > 0$, we can prove existence and uniqueness of solutions to the closed loop system. In this subsection, we derive the necessary conditions on gain parameters k_D and k_P so that given $\varepsilon > 0$ and $\Delta = 0$, we obtain $\limsup_{t\to\infty} |y(t)| \leq \varepsilon$.

Lemma 4.5 *Suppose the conditions in Lemma 4.4 are satisfied. Then*

$$\int_{t_0}^{t} (2aM(\tau) + b)u(\tau)\frac{dM}{d\tau}\, d\tau \geq 0.$$

Proof Define $g : \pi \to \mathbb{R}$ by $g(M(t)) := aM^2(t) + bM(t)$. As g is monotone increasing and differentiable on π and the hysteresis operator is output-rate dissipative, we have $\int_0^t ug'(M(t))dt \geq 0$, which proves the claim. □

Finally, we show that stabilization can be achieved by proper choice of parameters in the absence of disturbance Δ.

Theorem 4.3 *Suppose that the conditions in Lemma 4.4 are satisfied, and $\Delta = 0$. Then, for any $\varepsilon > 0$, there exists $\bar{k}_{P0} \geq k_{P0}$ such that, if $k_P > \bar{k}_{P0}$ and $\alpha k_D - k_P > 0$, then* $\limsup_{t\to\infty} |y(t)| \leq \varepsilon$.

Proof We employ a Lyapunov stability argument using La Salle's invariance principle. By Theorem 4.2, B is absolutely continuous. By Theorem 4.1, the operator $(I + \Gamma)^{-1}[\cdot; \psi_{-1}]$ is Lipschitz continuous on bounded subsets of $W^{1,1}(0, T)$. Hence, u is absolutely continuous on $(0, T)$ for any $T > 0$. Then, by Equation (9), we see that $W[f(x); \psi_{-1}]$ is absolutely continuous on $(0, T)$ for any $T > 0$. Equation (10) implies that $\dot{y}$ is locally absolutely continuous, that is, for almost every t, $\ddot{y}(t)$ exists.

Consider Equation (15) at time t,

$$\varphi(x)(t) + \beta u(t) = -\frac{k_P}{\alpha}\left(\dot{y} + \alpha y\right)(t) - \left(k_D - \frac{k_P}{\alpha}\right)\dot{y}(t). \tag{19}$$

By multiplying (19) with $\ddot{y} + \alpha\dot{y}$, and collecting terms, we find:

$$\begin{aligned}&\frac{k_P}{\alpha}\left(\dot{y} + \alpha y\right)\left(\ddot{y} + \alpha\dot{y}\right)(t) + \left(k_D - \frac{k_P}{\alpha}\right)\ddot{y}(t)\dot{y}(t)\\ &= \left(-\theta(x) - \beta u - \widehat{k}_P x\right)\left(\ddot{y} + \alpha\dot{y}\right)(t) - \left(\alpha k_D - k_P\right)\dot{y}^2(t).\end{aligned} \tag{20}$$

Consider the Lyapunov function

$$V(t) = \frac{k_P}{2\alpha}\left(\dot{y} + \alpha y\right)^2(t) + \left(k_D - \frac{k_P}{\alpha}\right)\frac{\dot{y}^2(t)}{2} + \beta\int_0^t (2aM + b)u\frac{dM}{d\tau}d\tau, \tag{21}$$

Then $V(t) \geq 0$. By (20) and (21), for almost every $t \in [0, \infty)$,

$$\dot{V}(t) = \left(-\theta(x) - \beta u - \widehat{k}_P x\right)\left(\ddot{y} + \alpha\dot{y}\right)(t) - \left(\alpha k_D - k_P\right)\dot{y}^2(t) + \beta\left(2aM(t) + b\right)u(t)\dot{M}(t). \tag{22}$$

From Equation (1), for almost every $t \in [0, \infty)$,

$$\left(\ddot{y} + \alpha\dot{y}\right)(t)\beta u(t) = \left(2aM(t) + b\right)\beta u(t)\dot{M}(t),$$

and
$$\left(\ddot{y} + \alpha\dot{y}\right)(t)\left(\theta(x)(t) + \widehat{k}_P x(t)\right) = \left(2aM(t) + b\right)\dot{M}(t)\left(\theta(x)(t) + \widehat{k}_P x(t)\right).$$

From Equation(22)

$$\dot{V}(t) = -\left(2aM(t) + b\right)\dot{M}(t)\left(\theta(x) + \widehat{k}_P x\right)(t) - \left(\alpha k_D - k_P\right)\dot{y}^2(t).$$

From Lemma 4.4, $2aM(t)+b > 0$ and, since $\dot{M}(t)$, $x(t)$, and $\theta(x)(t)$ have the same sign, $\dot{V}(t) \leq 0$. Thus $\dot{V}(t) \to 0$ as $t \to \infty$ for almost every t.

The invariant set for V is the set $(x, \dot{y}) = (0, 0)$. By Equation (7), the largest invariant subset of this set is given by:

$$\beta\, u = -k_P\, y.$$

Hence, for almost every t, $|y(t)| \leq \frac{\beta\,|u|}{k_P}$, which shows that for any $\varepsilon > 0$ we may choose $k_P \geq \bar{k}_P := \max\{k_{P0}, \frac{\beta\, d_1}{\varepsilon}\}$, where d_1 is the bound on u discussed in Lemma 4.4. Then, the largest invariant set for V will have the property: $|y| < \varepsilon$. □

5. Conclusion

In this paper, a novel control strategy for hysteretic systems associated with magnetic and smart actuators is proposed. The controller is a PD controller derived using two feedback signals. We proved regularity, well-posedness, and stability of the controller and then found conditions on the gain parameters for closed loop stability.

Journal of Physics: Conference Series **585** (2015) 012006 doi:10.1088/1742-6596/585/1/012006

6. References

[1] Dimitriadis E K, Fuller C R and Rogers C A 1991 Journal of vibration and acoustics, **113** (1) 100–107.
[2] Howe D 2000 Sensors and Actuators A: Physical **81** (1-3) 268-274.
[3] Baz A A, Ro J 1996 Smart Materials and Structures **5** (3) 135-144.
[4] Engdahl G 2000*Handbook of giant magnetostrictive materials* (Academic Press).
[5] Zhang T, Jiang C, Zhang H and Xu H 2004 Smart Material and Structures **13** (3) 473-477.
[6] Ekanayake D B, Iyer R V and Dayawansa W P 2007 American Control Conference 4321-4326.
[7] Ekanayake D B 2006 MS thesis (Texas Tech University, Lubbock, TX).
[8] Tan X and Baras J 2004 Automatica **40** (9) 1469–1480.
[9] Tan X and Baras J S 2005 IEEE Trans. Automatic Control **50** (6) 827–839.
[10] Song G, Zhao J, Zhou X and De Abreu-Garcia J A 2005 IEEE/ASME Trans. Mechatronics **10** (2) 198–209.
[11] Krejci P and Kuhnen K 2001 Proceedings of Control Theory and Applications IEEE **148** (3) 185–192.
[12] Tao G and Kokotovik P V 1995 IEEE Trans. on Automatic Control **40** (2) 200–212.
[13] Shieh H J, Lin F J, Huang P K, and Teng L T 2006 IEEE Trans. Industrial Electronics **53** (3) 905–914.
[14] Su C Y, Stepanenko Y, Svoboda J, and Leung T P 2000 IEEE Transaction on Automatic Control **45** (12) 2427–2432.
[15] Jayawardhana B, Logemann H, Ryan E P 2008 International Journal of Control **81** (8) 1331–1342.
[16] Valadkhan S, Morris K and Khajepour A 2010 International Journal of Robust and Nonlinear Control **20** (4) 460–471.
[17] Logemann H, Ryan E P and Shvartsman I 2007 ESAIM: Control, Optimization and Calculus of Variations **13** (3) 458–483.
[18] Valadkhan S, Morris K A and Khajepour A 2007 SIAM Journal of Applied Mathematics **67** (3) 667–686.
[19] Gorbet R B, Morris K A and Wang D W L 2001 IEEE Trans. Control Systems Technology **9** (1) 5–16.
[20] Ilchmann A, Ryan E P and Townsend P 2007 SIAM J. Control Optimization **46** (1) 210–230.
[21] Angeli D and Sontag E 2003 IEEE Trans. Automatic Control **48** (10) 1684–1699.
[22] Nealis J M and Smith R C 2007 IEEE Trans. Control Systems Technology **15** (1) 22–39.
[23] Ge P and Jouanech M 1996 IEEE Transactions on Control Systems Technology **4** (3) 209–216.
[24] Tan X and Iyer R V 2009 IEEE Control Systems Magazine **29** (1) 26–28.
[25] Cruz-Hernandez J M and Hayward V 2001 IEEE Trans. Control Systems Technology **9** (1) 17–26.
[26] Kuhnen K and Janocha H 1999 Proceedings of European Control Conference paper F 0291.
[27] Su C Y, Wang Q, Chen X and Rakheja S 2005 IEEE Transaction on Automatic Control **50** (12) 2069–2074.
[28] Iyer R V and Manservisi S 2006 Physica B **372**(1-2) 378–382.
[29] Iyer R V and Shirley M 2004 IEEE Transactions on Magnetics **40** (5) 3227–3239.
[30] Gorbet R B, Wang D W L and Morris K A 1998 Proceedings, IEEE international conference on robotics and automation **3** 2161–2167.
[31] Hughes D and Wen J T 1997 Smart materials and structures **6** (3) 287–300.
[32] Brokate M and Sprekels J 1996 *Hysteresis and Phase Transitions* Applied Mathematical Sciences Springer.

Journal of Physics: Conference Series **585** (2015) 012007 doi:10.1088/1742-6596/585/1/012007

The Hysteretic Behaviour of Partial Slip Elastic Contacts Undergoing a Fretting Loop

S Spinu, G Frunza

Department of Mechanics and Technologies, Stefan cel Mare University of Suceava, 13th University Street, 720229, Romania

E-mail: sergiu.spinu@fim.usv.ro

Abstract. Fretting defines an intrinsically multidisciplinary process, involving adhesion, oxidation, abrasion and pitting, in which mechanical contacts are subjected to alternating tangential displacements, small compared to dimensions of contact area. Assessment of hysteretic behaviour of stress and strain state in the contacting bodies is the main goal of this paper, which employs numerical analysis to extend the few existing analytical results to technologically important contact scenarios. An incremental iterative algorithm based on three levels of iterations, fully incorporating the interconnectivity between normal and tangential tractions, is reviewed in this work and used to simulate the hysteretic behaviour of partial slip elastic contacts undergoing fretting. Numerical simulations suggest that the fretting contact between dissimilarly elastic materials exhibits a unique path in the first two loading cycles, and then stabilizes with subsequent oscillating loading to a fixed trajectory, as in case of similarly elastic materials.

1. Introduction

Many industrial mechanisms involve mechanical contacts undergoing oscillating tangential displacements, small compared to dimensions of contact area. This type of loading conditions defines a fretting contact process. Related damages, i.e. fretting wear and fatigue, lead to a dramatic decrease of service lifetime of contacting machine elements. From a mechanical point of view, the problem is treated in the frame of Contact Mechanics, incorporating the Linear Theory of Elasticity in description of contacting bodies' responses to loading. Achievement of a closed-form description of the fretting contact between dissimilarly elastic materials is mainly impeded by the coupling between normal and tangential effects, the latter related to the capacity of the contact interface to sustain friction leading to development of shear tractions. A numerical study is therefore expected to advance the understanding of fretting phenomena and to provide assistance to the design of contacts with improved load-carrying capacity.

The method employed in this paper, also referred to as semi-analytical (SAM) [1], is based on superposition of fundamental solutions (i.e. Green functions) for the elastic half-space, applied within an iterative contact solver. The strong point of this approach is that domain discretization is limited to a surface region of the half-space boundary, imposed by trial-and-error as to enclose the actual contact area, as opposed to finite element analysis (FEA), in which a volume mesh of the entire bulk is required. It is asserted in [1] that SAM can be employed to solve a three-dimensional contact scenario with the computational complexity needed for FEA to simulate a two-dimensional contact process.

Study of gross-slip fretting contact using SAM originated in the works of Gallego, Nélias and Jacq [2], who solved repeatedly the contact model in the normal direction while accounting for the change in conformity due to wear. Chen and Wang [3] advanced a numerical model for the partial slip non-conforming contact considering tangential tractions, and predicted the effect of a tangential force, i.e. fretting mode I, in the contact between dissimilarly elastic materials. However, the loading history, related to the irreversibility of friction, was not accounted for. Further extensions addressed the problems of a supplementary torsional moment [4], i.e. fretting mode III, as well as the heterogeneity of elastic layered half-spaces [5]. Gallego, Nélias and Deyber [6] firstly implemented the incremental load application in simulation of the fretting contact, but the coarse temporal discretization impeded achievement of well-converged solutions, as proven by the jagged shape of the shear tractions profiles. Consequently, the study of hysteretic behaviour of fretting contacts was not properly conducted.

The main goal of this paper is to assess numerically the memory effect in the fretting contact between dissimilarly elastic materials. To this end, existing fast-converging algorithms [7,8] for the steady-state slip-stick contact, based on the Conjugate Gradient Method (CGM), are adapted to allow for transient conditions, and combined in an incremental iterative process, aiming to accurately reproduce the loading history in the contact process. The use of state-of-the-art numerical tools for convolution computation in the frequency domain, as the Discrete Convolution Fast Fourier Transform (DCFFT) technique [9], allows for simulations to be conducted on a fine spatial and temporal mesh, leading to well-converged numerical solutions.

2. Numerical modelling of fretting contact

The set of equations and inequalities describing the elastic contact are overviewed in this section, based on the works developed in [3,6,7,10]. The numerical formulation assumes all problem parameters are piece-wise constant on a $N_1 \times N_2$ uniformly spaced rectangular grid, also referred to as the computational domain A_P, established in the common plane of contact. The set A_P is expected to enclose the contact area A_C at any point on the loading curve. Problem spatial discretization is performed along directions $\vec{x}_1$ and $\vec{x}_2$ of a Cartesian coordinate system having its axes aligned with the grid sides. In this manner, analytical integration of arbitrary functions over arbitrary domains is circumvented and a numerical estimate is achieved through DCFFT assisted multi-summation. Although a purely elastic analysis is intended in this paper, due to the irreversibility of friction, the state achieved at any point on the loading curve is expected to depend on all previous states. Consequently, discretization of the loading path, i.e. in the temporal domain, is required as well, implying that load should be applied in small increments, and the contribution of each increment should be superimposed to previous levels in order to acquire a good assessment of the current state. However, no rate-dependence is considered in this model, meaning no impact study is conducted.

In the numerical formulation, it is convenient to denote all contact parameters as functions of integers indexing the grid cells, $i=\overline{1,N_1}$, $j=\overline{1,N_2}$, as well as of the number of imposed loading steps, $k=\overline{1,N_\ell}$, e.g. $p^{(k)}(i,j)$ denotes the nodal pressure at the intersection of the line i with the column j of the rectangular grid, achieved after application of k loading increments. With this notation convention, the well-known equations governing the elastic contact in the normal direction yield:

$$W^{(k)} = \Delta \sum_{(i,j)\in A_C^{(k)}} p^{(k)}(i,j)\,; \tag{1}$$

$$h^{(k)}(i,j) = hi(i,j) + u_3(i,j,k) - \omega_3^{(k)},\ (i,j)\in A_P\,; \tag{2}$$

$$p^{(k)}(i,j) > 0 \quad \text{and} \quad h^{(k)}(i,j) = 0,\quad (i,j)\in A_C^{(k)}\,; \tag{3}$$

$$p^{(k)}(i,j) = 0 \quad \text{and} \quad h^{(k)}(i,j) > 0,\quad (i,j)\in A_P - A_C^{(k)}\,. \tag{4}$$

Equation (1) expresses the static force equilibrium, relating the normal force W to the induced pressure distribution p. Digitization of geometrical condition of deformation yields the interference equation (2), in which the gap between deformed surfaces h is expressed in terms of initial gap (or contact geometry) hi, of surface normal displacement u_3, and of rigid-body translation ω_3. The contact or non-contact status of every cell in the A_P is assessed by the complementarity contact conditions in equations (3) and (4). An analogous model can be established for the slip-stick elastic contact in the tangential direction, provided a torsional moment is not present:

$$T_n^{(k)} = \Delta \sum_{(i,j)\in A_C^{(k)}} q_n^{(k)}(i,j),\ n=1,2; \tag{5}$$

$$\begin{bmatrix} s_1^{(k)}(i,j)-s_1^{(k-1)}(i,j) \\ s_2^{(k)}(i,j)-s_2^{(k-1)}(i,j) \end{bmatrix} = \begin{bmatrix} u_1^{(k)}(i,j)-u_1^{(k-1)}(i,j) \\ u_2^{(k)}(i,j)-u_2^{(k-1)}(i,j) \end{bmatrix} - \begin{bmatrix} \omega_1^{(k)}-\omega_1^{(k-1)} \\ \omega_2^{(k)}-\omega_2^{(k-1)} \end{bmatrix},\quad (i,j)\in A_C^{(k)}; \tag{6}$$

$$\left\|\mathbf{q}^{(k)}(i,j)\right\| \le \mu p^{(k)}(i,j) \quad \text{and} \quad \left\|\mathbf{s}^{(k)}(i,j)-\mathbf{s}^{(k-1)}(i,j)\right\| = 0,\ (i,j)\in A_S^{(k)}; \tag{7}$$

$$\left\|\mathbf{q}^{(k)}(i,j)\right\| = \mu p^{(k)}(i,j) \quad \text{and} \quad \left\|\mathbf{s}^{(k)}(i,j)-\mathbf{s}^{(k-1)}(i,j)\right\| > 0,\ \ (i,j)\in A_C^{(k)} - A_S^{(k)}, \tag{8}$$

where $\mathbf{T}(T_1,T_2)$ is the tangential force, $\mathbf{q}(q_1,q_2)$ the vectors of shear tractions, $\mathbf{s}(s_1,s_2)$ the relative slip distances, μ the frictional coefficient, and A_S the stick area. The torsional moment, which adds a supplementary unknown, namely the rotation angle along direction of $\vec{x}_3$, but also an additional moment equation in the static equilibrium, should always be considered in simulation of fretting mode III. However, this analysis is restricted for brevity to fretting modes I and II. The model (5) – (8) also assumes the tangential force is not large enough to induce gross-slip, which is guaranteed at any point on the loading curve if $\mathbf{T}^{(k)}$ is smaller than the instantaneous limiting friction force $\mu W^{(k)}$. On the other hand, both Cattaneo [11] and Mindlin [12] proved independently that, even when the contacting bodies are globally sticking, a peripheral region of slip is required to relinquish the infinite shear tractions at the boundary of the contact area, or else the no-slip contact scenario is not compatible with either Linear Theory of Elasticity or Coulomb's law of friction.

Providing a subsequent tangential load increment is applied in a slip-stick contact, the stick or slip status of specific points on the contact area is expected to change, as predicted by the complementarity conditions in equations (7) and (8). Points included in the stick zone conserve their status if and only if no relative motion occurs between matching particles on the two contacting surfaces, i.e. the increment of relative slip distance in the considered timeframe is nil. Points for which this condition is not met pass into the slip zone. As pointed out by Hills, Nowell, and Sackfield [13], the existence of slip is intrinsically conditioned by an increase or decrease in the level of tangential load, and therefore a purely static model is not appropriate. Moreover, when the tangential load increment changes sign, the rate of change of shear tractions and of relative tangential displacements also changes sign. As shear tractions no longer oppose relative motion between the contacting bodies, instantaneous stick occurs, and further loading leads to partial slip in the reversed direction. Subsequent state can be computed starting from a fully-stuck one, i.e. with vanishing relative slip distances, but with an initial shear tractions distribution. The model advanced herein is consistent with this formulation, provided the point on the loading curve when the tangential load trend is reversed is accurately reproduced.

Considering the similarity between the set of equations (1) – (4) and (5) – (8), the same type of algorithm could be used to solve either model considered independently. The main difficulty consists in the system being intrinsically non-linear due to presence of rigid-body translations. Also, its size is not known in advance, as neither contact nor stick areas are known a priori. In order to use the CGM, which is proven to converge only for systems with symmetric and positive definite matrix, the static equilibrium is imposed in a correction of the system solution outside the CGM core, as depicted in figure 1, and estimates for the rigid-body translations are computed numerically from the interference equation, thus linearizing the system. This scheme aims to preserve the influence coefficients matrix, which is intrinsically symmetric and positive definite, as the system matrix. Detailed description of the algorithms can be found in [7] for the set of equations (1) – (4), and in [8] for the set (5) – (8).

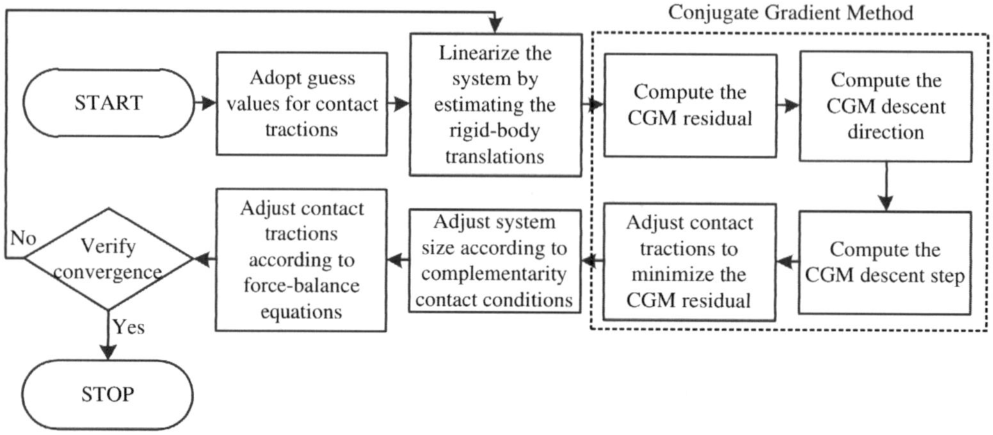

Figure 1. A CGM based scheme to compute contact tractions.

Moreover, when materials of contacting bodies are dissimilarly elastic, the two set of equations are coupled, i.e. resolution of each requires solution of the other, as elastic displacement result by superposing contributions of all contact tractions, both normal and shear. To surmount this mutual determination, an iterative approach, fully described in [14], is employed, allowing simulation of the elastic three-dimensional fretting contact between dissimilarly elastic materials with the additional cost of a supplementary outer loop.

3. Results and discussions

The spherical contact between similarly or dissimilarly elastic materials, subjected to fretting, is simulated using the proposed computer program. A ball of radius $R = 18\text{ mm}$ is pressed against an elastic half-space following the loading path depicted in figure 2. The maximum level of W is fixed in all simulations, while the maximum tangential force T_{lim} is chosen in relation to the imposed frictional coefficient, i.e. $T_{\text{lim}} = 0.9\mu W$, assuring a gross-slip regime is not reached. The load is imposed in a variable number of increments, and the effect upon solution convergence is discussed.

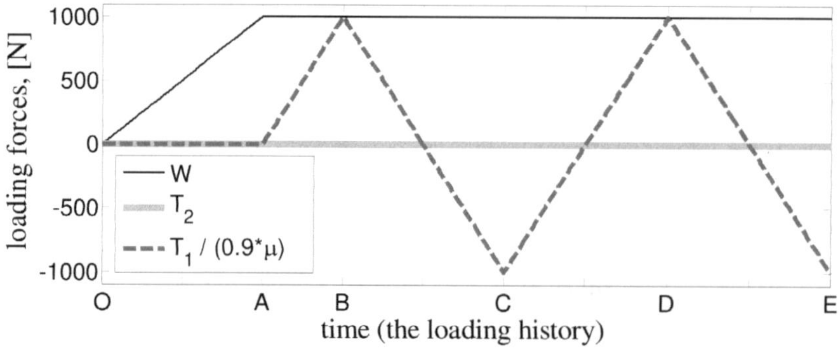

Figure 2. The loading path imposed in the fretting contact simulation.

When the contacting materials are assumed similarly elastic (Young modulus $E = 210\,\mathrm{GPa}$, Poisson's ratio $\nu = 0.3$), the Hertz frictionless theory predicts a central pressure $p_H = 1.996\,\mathrm{GPa}$ and a contact radius $a_H = 0.489\,\mathrm{mm}$, which are used as normalizers in figures 3-4, together with the rigid-body tangential displacement $\omega_{\lim}$ corresponding to $T_{\lim}$. Due to decoupling between normal and tangential effects, any state on the loading trajectory OA can be obtained in one step, by running one solver iteration consisting in resolution of contact problem in the normal direction. As no shear force is explicitly applied, the friction related memory effect is not present. When an oscillating tangential force is subsequently applied (point A in figure 2), shear tractions profiles on the loading path ABCD, corresponding to equal loading levels $T_1 = 0$ and $T_1 = T_{\lim}/2$ but laying on different trajectories, prove that the current state depends not only on the achieved loading level, but also on the loading history, as depicted in figure 3. However, provided the states when the rate of change in the tangential load changes sign are accurately captured by a suitable choice of the loading increments, no additional temporal discretization is required. For example, any state on the loading trajectory BC can be accurately predicted in only two loading steps, the first imposing the load corresponding to point B, and the second the currently achieved loading level.

The numerical predictions for this contact scenario allow validating the computer program against closed-form solutions reviewed by Johnson [10]. The numerical data, depicted using discrete symbols in figure 3, match well its analytical counterpart, traced as continuous line. As depicted in figure 4, the force – rigid-body displacement curve exhibits a hysteretic behaviour, also referred to as a fretting loop. The closed-form description of this contact scenario predicts that the loop closes perfectly, implying states corresponding to points B and D on the loading curve overlap, which is also obtained through numerical simulation within an imposed precision. A unique path is thus established from the simulation of trajectory BCD. In simulation of this path, one can take advantage of the feature of symmetry, i.e. states on the CD trajectory are a complete reversal of those on BC.

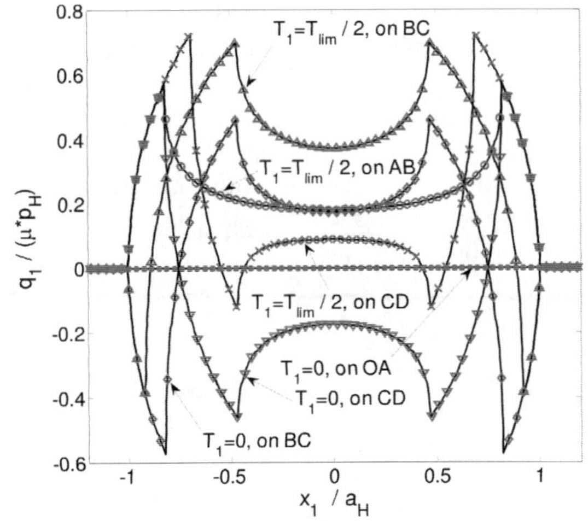

Figure 3. Shear tractions profiles in the plane $x_2 = 0$.

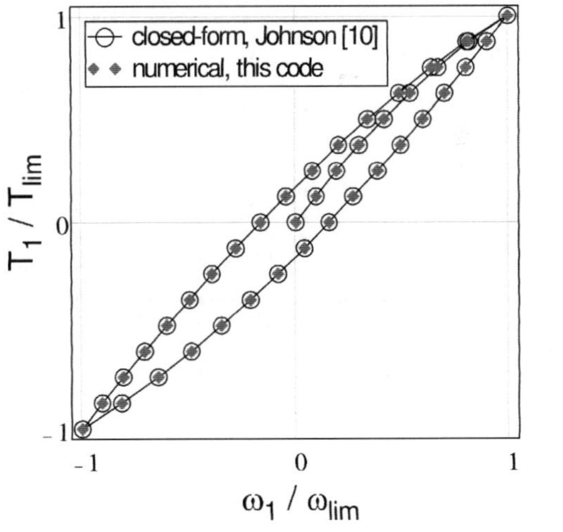

Figure 4. A fretting loop in the contact between similarly elastic materials.

A different behaviour is predicted numerically when the normal and tangential effects are coupled. A dissimilarly elastic contact scenario, for which, to our best knowledge, no analytical solution has been advanced, is subsequently simulated using the advanced algorithm. To this end, the ball in the previous contact scenario is considered rigid. The Hertz frictionless framework predicts that $p_H = 3.168\,\mathrm{GPa}$ and $a_H = 0.388\,\mathrm{mm}$, which are used as normalizers in figures 5-7.

Journal of Physics: Conference Series **585** (2015) 012007 doi:10.1088/1742-6596/585/1/012007

On the OA loading trajectory, although no tangential force is explicitly applied, divergent pressure induced tangential displacements of initially matching surface points arise due to dissimilarity in elastic responses of contacting materials. The resulting relative tangential motion is resisted by friction, leading to development of shear tractions. It should be noted that tangential displacements due to pressure also arise in the contact between similarly elastic materials, but of the same magnitude and in the same direction, thus leading to no relative motion and consequently to vanishing shear tractions.

The frictional contact process which is established in case of dissimilar contacting materials is found to be very sensitive to the temporal discretization. As also stated in [6], large loading increments predict jagged shear tractions profiles, due to important contractions / dilatations of the stick zone following application of subsequent loading increments. In this simulation, the loading process was divided into 900 increments (200 load steps for the OA trajectory, and 700 for remaining of the loading path in figure 3), leading to well-converged solutions.

To further prove the need for incremental load application in simulation of frictional contact processes when dissimilarly elastic materials are involved, a closed-form relation derived by Spence [15] based on the condition that shear stresses should be continuous over the boundary between the slip and the stick regions is used as benchmark:

$$\frac{\mu}{\beta} = \left[\ln\frac{1+\bar{a}_S}{1-\bar{a}_S}\cdot\left(2\bar{a}_S K\sqrt{1-\bar{a}_S^2}\right)\right]^{-1}, \tag{9}$$

where a_S is the stick radius, $\bar{a}_S = a_S/a_H$, K the complete elliptic integral of the first kind, and β the Dundurs constant [16], assessing the dissimilarity in elastic properties of contacting materials ($\beta = 0.286$ in the considered contact scenario). The stick radii predicted numerically with or without simulation of the loading history (i.e. when load is applied incrementally or in one step, respectively) are compared in figure 5 to the analytical curve (9). Although Spence's result is derived under the Goodman approximation [17], the agreement is considered satisfactory giving the model complexity. The gap to the analytical curve when $\bar{a}_S \to 1$ proves the inability of the numerical approach to handle asymptotic behaviours. Indeed, in the discretized formulation, the stick region can only vary in increments equal to the area Δ of the elementary patch of the imposed grid. Consequently, the finer the grid, the smaller the slip zone that can be predicted, but also the greater the computational effort required per iteration. When the never vanishing slip region decreases toward zero, extents of the stick area may therefore be underestimated in the numerical solution due to inherent discretization errors. However, predictions obtained without properly simulating the loading history may be unreliable, as depicted in figure 5, and should be considered with great care for designing purposes.

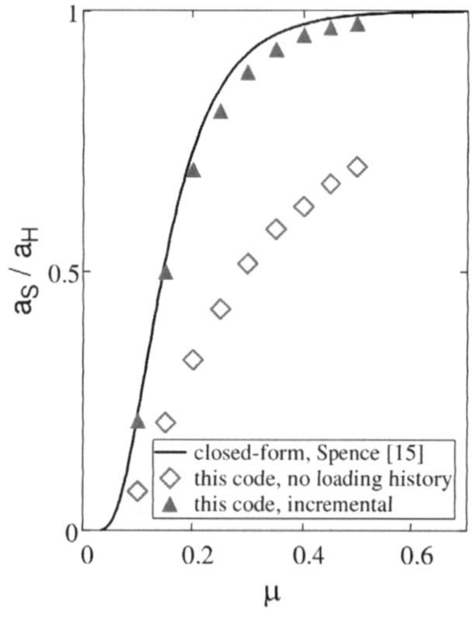

Figure 5. Comparison of existing closed-form expression with numerical predictions obtained with or without proper simulation of loading history in the indentation (fretting mode II) involving dissimilarly elastic materials.

The load - displacement curves for various frictional coefficients depicted in figure 6 predict that the fretting loop does not closes perfectly, as in case of similarly elastic materials, but a stabilized regime emerges past point E on the loading curve, when the tangential force reaches the value $-T_{\lim}$ for the second time. This assertion is also supported by the shear tractions profiles depicted in figure 7 for $\mu = 0.6$, although the same tendency was also predicted for all simulated frictional coefficients values ranging between 0.1 and 0.6. However, the shear tractions profiles were found to vary from one case to another, making the achievement of a global description improbable. While states corresponding to points B and D on the loading curve differ significantly, as depicted in figure 7, those matching points C and E overlap within an imposed precision. It was verified by continuing the simulation on an extended loading path that further periodical loading leads to states superimposing the loop established in the contact process between points C and E. A similar periodic stability was numerically found in [5] for the fretting contact of layered half-spaces.

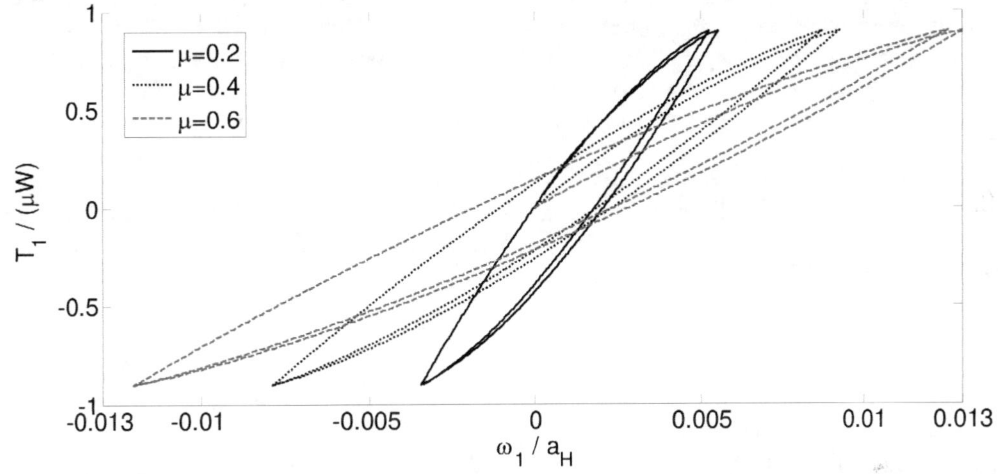

Figure 6. The fretting loop during the first two loading cycles in the contact between dissimilarly elastic materials subjected to oscillating tangential displacement (fretting mode I).

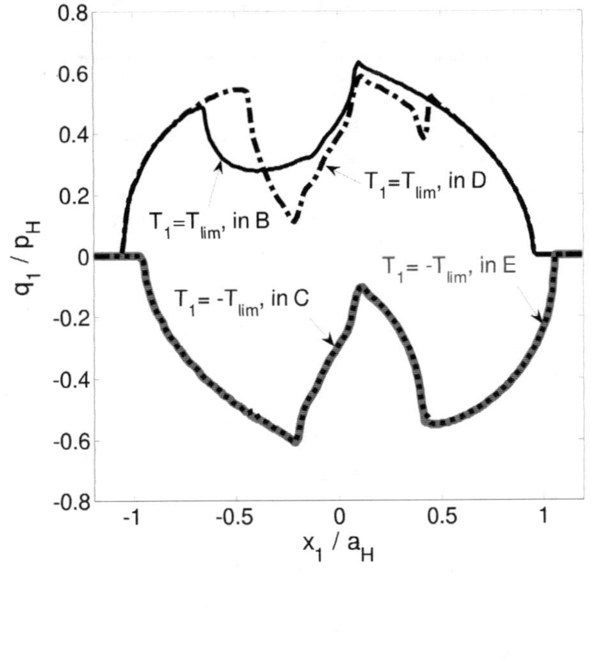

Figure 7. Achievement of a periodic stability proven by matching shear tractions profiles in the first two loading cycles of the fretting contact between dissimilarly elastic materials, $\mu = 0.6$.

Journal of Physics: Conference Series **585** (2015) 012007 doi:10.1088/1742-6596/585/1/012007

4. Conclusions

A numerical solution for the three-dimensional slip-stick elastic contact between dissimilarly elastic materials is employed in this paper to predict the hysteretic behaviour of the fretting contact, related to the irreversibility of friction.

Simulation of fretting contact between similarly elastic materials allows for program validation, and proves that any state on the fretting loop, which is established as a fixed path from the first loading cycle, can be predicted with a minimum number of increments, related to the number of changes in the monotonicity of the tangential load.

Further validation is achieved in case of indentation involving dissimilarly elastic materials (fretting mode II), when shear stresses arise due to relative motion induced by the dissimilarity in elastic response of contacting materials. To obtain well-converged numerical solutions, a fine discretization in the temporal domain is critical.

When an oscillating tangential load is subsequently applied (fretting mode I), the contact of dissimilarly elastic materials exhibits a unique trajectory in the first two loading cycles, then stabilizes to a steady path, as in case of similarly elastic materials. The knowledge of this path and of corresponding contact tractions and stresses is essential in prediction of yield inception or crack nucleation in the mechanical contact, providing grounds for a better understanding of premature contact failure due to fretting wear or fretting fatigue.

Acknowledgments

This paper was supported by the project "Progress and development through post-doctoral research and innovation in engineering and applied sciences – PRiDE - Contract no. POSDRU/89/1.5/S/57083", project co-funded from European Social Fund through Sectorial Operational Program Human Resources 2007-2013.

References

[1] Renouf M, Massi F, Fillot N and Saulot A 2011 *Tribol. Int.* **44** p 834
[2] Gallego L, Nélias D and Jacq C 2006 *ASME J. Tribol.* **128** p 476
[3] Chen W W and Wang Q J 2008 *Mech. Mater.* **40** p 936
[4] Wang Z J, Meng F M, Xiao J X and Wang W Z 2011 *P. I. Mech. Eng. J.-J. Eng.* **225** p 72
[5] Wang Z J, Wang W Z, Wang H, Zhu H and Hu Y Z 2010 *ASME J. Tribol.* **132** 021403
[6] Gallego L, Nélias D and Deyber S 2010 *Wear* **268** p 208
[7] Polonsky I A and Keer L M 1999 *Wear* **231** p 206
[8] Spinu S and Frunza G 2011 *Proc. VI Int. Conf. Balttrib (Kaunas, Lithuania, 17-19 November 2011)* (Kaunas: Aleksandras Stulginskis University) p 105
[9] Liu S B, Wang Q and Liu G 2000 *Wear* **243** p 101
[10] Johnson K L 1985 *Contact Mechanics* (Cambridge: University Press)
[11] Cattaneo C 1938 Sul contatto di due corpi elastici: distribuzione locale degli sforzi *Accademia Nazionale Lincei Rendiconti Ser. 6* **27** (Roma: L'Accademia) pp 342-348, 434-436, 474-478
[12] Mindlin R D 1949 *ASME J. Appl. Mech.* **16** p 259
[13] Hills D A, Nowell D and Sackfield A 1993 *Mechanics of elastic contacts* (Oxford: Butterworth-Heinemann) pp 111-118
[14] Spinu S and Glovnea M 2012 *J. Balk. Tribol. Assoc.* **18** p 195
[15] Spence D 1975 *J. Elasticity* **5** p 297
[16] Dundurs J and Lee M S 1972 *J. Elasticity* **2** p 109
[17] Goodman L E 1962 *ASME J. Appl. Mech.* **29** p 515

Method of Integration for Equation of Two Balls in Dumped Collision

S Alaci[1], D A Cerlincă[1] , F C Ciornei[1], C Filote[2] and G Frunză[1]

[1] Mechanics and technology Department, University of Suceava, 13 Universitatii Street, 720229 Suceava, Romania
[2] Faculty of Electrical Engineerig and Computer Science, University of Suceava, 13 Universitatii Street, 720229 Suceava, Romania

E-mail: stalaci@yahoo.com

Index terms: experimental mechanics, rolling hysteresis, photoelasticity

Abstract. The paper presents an integration method for the equation characteristic to a dumped collision. The equation which describes the dumped collision is nonlinear ODE of second order. The solution consists from two stages: in the first stage the equation order is reduced with a unity finding a prime integral. In this moment the maximum approach can be evaluated. Next, using a elementary numerical method the relation between approaching versus time is found. The method is applied for two equations which describe a dumped collision. The results obtained with the proposed method are in a perfect agreement with those from original paper. Finally, in order to emphasize the difference between the two models and the effect of dumping, the hysteresis loops for two values of the coefficient of restitution, corresponding to the two models are plotted to emphasize the difference are plotted.

1. Introduction

In multibody dynamics, collision or impact phenomena appear when velocities are changed suddenly. The collisions can be studied considering that the impact is instantaneously, using coefficients of restitution (COR) on one hand [1], or, considering that during impact process the kinematical and dynamical parameters present continuous variation, on the other hand.

The method considering the instantaneous character of impact has the main advantage of straightforward calculus, reduced in fact to algebra, but presents the essential inconvenience that cannot estimate the magnitude of forces occurring during collision. Additionally, a problem concerning the definition modality for the coefficient of restitution, (COR), arises. More precise, the technical literature mentions two modalities of defining the coefficient of restitution, [2], [3]. The first method is due to Newton and considers the kinematical definition of coefficient of restitution as the ratio, with opposite sign, of normal components of relative velocities of the initial contact points, after and before collision, respectively. Taking this modality of defining COR can lead to paradoxical situations. Such a circumstance is was exposed by Kane, [4], who presents the case of collision with friction of a mechanical system for which, the conclusion is that kinetical energy of the system is greater at final stage than initially. To eliminate this inconsistency, is required the acceptance of another hypothesis, specifically: the collision happens during a finite period of time, being characterised by two phases - the approaching and detaching phases. In this situation, the coefficient of restitution is defined according to Poisson as the ration between the percussion forces corresponding to detaching and approaching, respectively.

The challenge of describing the impact in a general form, considering all the parameters influencing the process, is practically impossible. From this cause, most of the works concerning the study of collisions treat systems considering more or less simplifying hypothesis. The simplest model met in elementary dynamics works is for the collision of two punctiform bodies. The next model regarding problem complexity considers centric collision of two spheres made of elastic, homogenous and isotropic materials.

Goldsmith, [5], presents a solution due to Timoshenko, [6], for a perfect elastic impact between two balls of different radii. The solution found by Timoshenko is based on the relation force-approach for the case of a system of two elastic balls. For this model Timoshenko found the analytical relation for maximum approaching and for the contact time. For the case of a more complicated model, that is the dumped collision, the problem becomes considerably complex. The simplest model describing the behaviour of this collision model is the Kelvin-Voigt model which was used for the dynamic system. Hunt and Crossley [7] showed that the model has the disadvantage of an open hysteresis loop and for the final stage of the impact process the two bodies attract each other instead of rejecting. They indicated that in order to obtain a closed hysteretic loop it is necessary to use a modified Kelvin-Voigt model with variable damping and elastic coefficients.

Lankarani and Nikravesh, [8], following the results of Hunt and Crossley, [7], obtained the equation for centric impact between two elastic balls.

2. Lankarani Nikravesh Model

The collision with dumping of two balls, as sketched in figure 1, is considered.

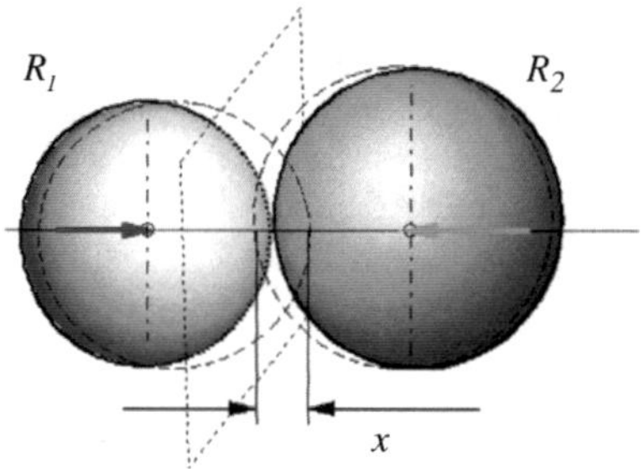

Figure 1. Two colliding balls.

The equation propose by Lankarani and Nikravesh, was obtained based, as in Timshenko's solution, on the force-displacement relation from the Hertz's problem to which was added the hypothesis that the dumping force during collision depends both on the approaching velocity of the bodies and on a power of bodies approaching. The tangible form of the equation obtained by the two authors is, [8]:

$$F = Kx^{\alpha}\left[1 + \frac{3(1-e^2)}{4}\frac{\dot{x}}{v_0}\right] \tag{1}$$

where:

e is coefficient of restitution defined by Newton as the ratio of the normal components of relative velocities of initial contact points for the detaching and approaching phase, respectively;

x is the distance between the initial contacting points:

v_0 is initial impact velocity;

α is an exponent which for point contact is $\alpha = 3/2$;

K is a coefficient characterising the elastic proprieties and the geometry of the balls around the contact points. More exactly,

$$K = \frac{4}{3(\eta_1 + \eta_2)}\left[\frac{R_1 R_2}{R_1 + R_2}\right]^{1/2} \tag{2}$$

and

$$\eta_{1,2} = \frac{1 - \nu_{1,2}^2}{E_{1,2}} \tag{3}$$

Where $R_{1,2}$ are the balls radii, $E_{1,2}$, $\nu_{1,2}$ the Young modulus and Poisson coefficients, respectively.

The equation 1 is a ordinary nonlinear differential equation. The equation has no analytic solutions. In the paper the authors don't give any information about the methodology of integration of this equation. They present only the results of integration under graphical form which show the influence of parameters e and v_0 upon the kinematics and dynamical parameters and the forms of hysteresis loops. A key advantage of the model consists in presenting closed hysteresis loops closed in the origin.

3. Proposed method of integration

The authors suggest a method of integration for the equation (1), made in two steps:

1. The first step consist in finding an analytical relation between the normal approach and velocities:
2. The second step concerns numerical computing of time dependence of approach.

For the first step we use the method of the Horway and Veluswami, [9]. We write the equation (1) as:

$$\ddot{x}(t) + \gamma\, x(t)^n \dot{x}(t) + \chi\, x(t)^n = 0 \tag{4}$$

where upper dot represents time derivative. Using the notations:

$$M = \frac{m_1 m_2}{m_1 + m_2} \tag{5}$$

where M the reduced mass of the system (are the masses of the balls) and :

$$\chi = K / M\ ; \quad \gamma = K\frac{3(1 - e^2)}{4v_0} \tag{6}$$

$$\frac{dv}{dx}\frac{dx}{dt} + \gamma\, x^{3/2} v + \chi\, x^{3/2} = 0 \tag{7}$$

$$\frac{dv}{dx} v + \gamma\, x^{3/2} v + \chi\, x^{3/2} = 0 \tag{8}$$

$$\chi\, x^{3/2} = -\frac{v dv}{1 + \frac{\chi}{\gamma} v} \tag{9}$$

The general solution of the above equation is:

$$\frac{2}{5}\chi x^{5/2}+C=\frac{\chi}{\gamma}v-\frac{\chi^2}{\gamma^2}ln(\chi+\gamma v) \tag{10}$$

In order to find a particular solution the initial conditions is imposed. At moment $t=0$ we need $x=0,\ v=v_0$. So, for the approaching phase the equation (10) becomes:

$$\frac{2}{5}\chi x^{5/2}=-\frac{\chi}{\gamma}(v-v_0)+\frac{\chi^2}{\gamma^2}ln\frac{\chi+\gamma v}{\chi+\gamma v_0} \tag{11}$$

Denoting t_m as the moment of reaching maximum approach x_m, from condition that at the moment $t=t_m$, $x=x_m$ and $v=0$ we can find the maximum approach;

$$x_m=\left\{\frac{5}{2}\left[\frac{1}{\gamma}v_0-\frac{\gamma}{\chi}ln\left(1+\frac{\gamma}{\chi}v_0\right)\right]\right\}^{5/2} \tag{12}$$

For the detaching phase, the conditions imposed at $t=t_m$ are: $v=0$, $x=x_m$ and conduct to the solution form for the coming off phase.

$$-\frac{2}{5}\chi\left[x'^{5/2}\right]-x_m^{5/2}=\frac{\chi}{\gamma}v-\frac{\chi^2}{\gamma^2}ln\left(1+\frac{\gamma}{\chi}v\right) \tag{13}$$

In order to find the time dependence of x on the approach interval we consider that the equation (11) give the dependence of v as function of x.. This equation is a transcendental one and must be solved numerically. There are many software applications which can resolve it, most of these being based on the Runge-Kutta, [10] algorithm. In order to accurately describe the impact phenomenon, a large number of points should be considered. In order to reduce the computing time, a simple program was created for solving the equation by the bipartition method. Applying the Runge-Kutta method assumes that the functions are derivable and require much time compared to the bipartition method. Different forms of the relationship between velocity and displacement are obtained for the compression and expansion phases. The relations (11) and (13) are expressed under the form:

$$f(v,x)=\frac{2}{5}\chi x^{5/2}+\frac{\chi}{\gamma}(v-v_0)+\frac{\chi^2}{\gamma^2}ln\left(\frac{\chi+\gamma v_0}{\chi+\gamma v}\right)=0 \tag{14}$$

where x increases from zero to x_m

$$g(v',x')=\frac{2}{5}\chi\left[x'^{5/2}-x_m^{5/2}\right]+\frac{\chi}{\gamma}v'-\frac{\chi^2}{\gamma^2}ln\left(1+\frac{\gamma}{\chi}v'\right)=0 \tag{15}$$

and x' decreases from x_m to zero.

For the interval $[0,x_m]$ we choice a set of equidistant points $x_k, k=0\div n$, and for every x_k we solve the equation (14) using bipartition method, with accuracy of 10^{-15}. So we obtained n pairs, (x_k,v_k). Similarly, for equidistant points x'_k for the detaching phase, pairs of points are obtained after numerical solving the equations (14) and (15). The velocity-displacement dependence is represented in figure 2, foe equidistant points from the range. As it can be noticed, in the vicinity of the point of maximum indentation, the points of the plot are more remote. To diminish this effect, it is necessary to choose the sequences x_k, x'_k so that the points should be denser in the vicinity of point x_m. To this purpose, the points are dispersed inside the interval $[0,x_m]$ obeying the following law:

$$x_k = \frac{1-\alpha^{k/n}}{1-\alpha},\ 0<\alpha<1;\quad x'_k = \frac{\beta^{k/n}-\beta}{1-\beta} \tag{16}$$

where $\beta = 1/\alpha$. For $COR = 0.999$, $n = 20$ and $\alpha = 1/200$, the pairs of points (x_k, v_k) and (x'_k, v'_k) are represented in figure 3.

One can observe that the points of the plots are more compact in the vicinity of the point x_m.

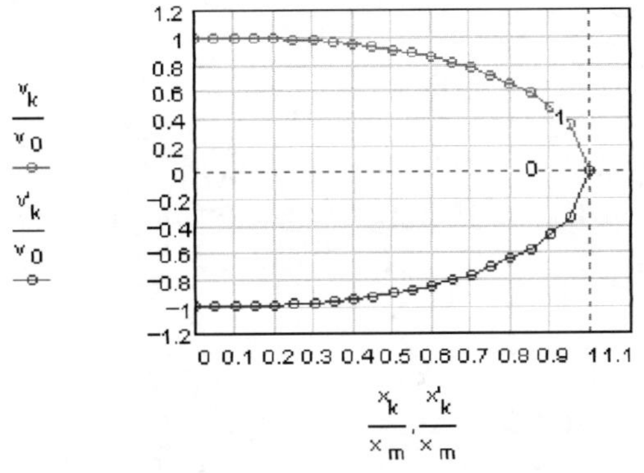

Figure 2. Velocity-displacement dependence for equidistant points

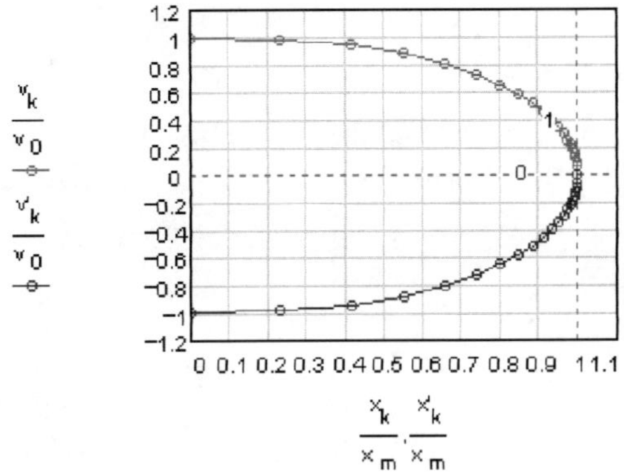

Figure 3. Velocity-displacement dependence for proposed method

The equation:

$$v = dx/dt \tag{17}$$

can now be integrated numerically, using Euler method. Therefore:

$$\begin{aligned} &for\ k=0,\ t_0=0, \\ &for\ k=1:n-1,\ t_k = t_{k-1} + \frac{x_k - x_{k-1}}{v_k} \end{aligned} \tag{18}$$

In equation (18) $k \neq n$ because $v_n \equiv 0$. The approaching time t_m cannot be obtained using (18). Plotting the dependence $x_k = x_k(t_k), k = 0 \div n-1$, we obtained the curve plotted in figure 4, and it can be seen that the elapsed time between the moments correspondent to successive pairs (x_k, v_k) increases as the moment of maximum indentation value x_m is closer.

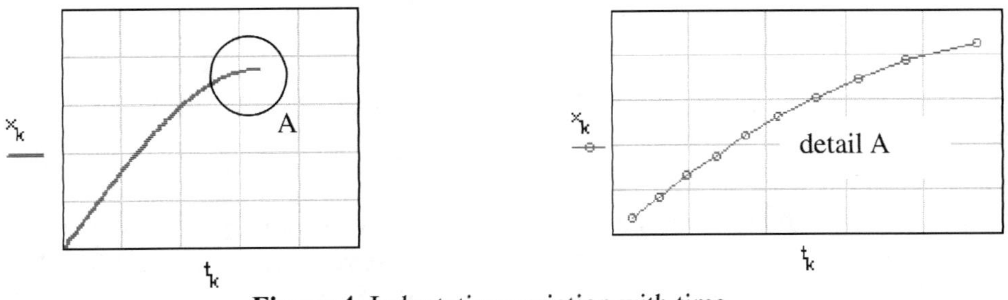

Figure 4. Indentation variation with time

The problem of finding the approaching time $t_m = t_n$ is important as the indentation time is one of the characteristic parameters of the impact process. In order to find the time t_n, the curve from Fig. 4 must be extrapolated. The shape of the curve from figure 4 suggested the next extrapolation possibilities.

1. The dependence $x = x(t)$ is approximated by a sinusoid having the form $x = A\,sin(\mu t)$, where the unknowns A and μ are found from the condition of minimum for the function:

$$S(A,\mu) = \sum_{k=0}^{n-1} [(A\,sin(\mu t_k) - x_k]^2 \tag{19}$$

By t'_m is denoted the value found for the approaching time.

2. The dependence $x = x(t)$ is approximated by a sinusoid having the form $x = A\,sin(\mu t)$ but, from the beginning, the amplitude is imposed to x_n and from the condition of minimum of the function:

$$T(\mu') = \sum_{k=0}^{n-1} \left[\frac{x_m}{sin(\mu t''_m)} sin(\mu' t_k) - x_k \right]^2 \tag{20}$$

only the value of parameter μ' remains be found. In this case, the approaching time found is denoted by t''_m.

3. It is imposed the condition that the center of the circle passing through the last three points of the plot should be positioned on the vertical of the last point. Let the coordinates of the center of the circle be (t_O, x_O). Practically, when the three points are very closed, the circumscribed circle is the osculator circle at the plot $x = x(t)$ in the point (t_n, x_n).

$$\begin{cases} (t_{n-2} - t_O)^2 + (x_{n-2} - x_O)^2 = r^2 \\ (t_{n-1} - t_O)^2 + (x_{n-1} - x_O)^2 = r^2 \\ ((t_n - t_O)^2 + (x_n - x_O)^2 = r^2 \\ t_O = t_n \end{cases} \tag{21}$$

This last value for the approaching time is denoted t'''.

In order to decide which of the above three values of compressing time must be considered, it is accepted the assumption that that for values of coefficient of restitution closed to unity, the collision can be considered elastic. For the elastic collision there are analytical relations both for maximum approach and for compressing end expansion times, times having identical values for the elastic impact. Goldsmith provided the following relations:

$$t_{el} = \frac{x_{el}}{v_0} \int_0^1 \frac{1}{\sqrt{1 - x^{5/2}}} dx, \; x_{el} = \left[\frac{5Mv_0^2}{4K} \right]^{2/5} \tag{22}$$

The tests for different values of COR close to unity and for a division of approaching distance in 20,000 points proved that the closest value to t_{el} is the one obtained with the third hypothesis. For a value for coefficient of restitution $e = 0.999$, for the three alternatives of extrapolation, the following results were obtained: $t'/t_{el} = 1.041884$, $t''/t_{el} = 1.024045$, $t'''/t_{el} = 0.999574$. By reporting to t_{el} the times obtained, the dimensionless time obtained facilitates the comparison and thus, the third alternative $t_m \equiv t'''$ was chosen.

For the detaching phase, the equation can be integrated without difficulty using Euler method.

$$for\ k=0,\ t'_0=t_m,$$
$$for\ k=1:n,\ t'_k=t'_{k-1}+\frac{x'_k-x'_{k-1}}{v'_k} \tag{23}$$

The Lankarani-Nikravesh model has the disadvantage, as the authors mention, of being appropriate only for quasielastic collision, for $e>0.85$. The attempt of applying the model for the cases of smaller COR, $e<0.85$, leads to contradictory aspects. The coefficient of restitution found based on relation (1), $e_{out}=v_f/v_0$, where v_f is the relative velocity of contacting points for complete detaching, is greater than the initial coefficient and the difference between these two values increases with increasing internal friction.

Recently, Flores, [11] reached the conclusion that the Poincarè map of the model must be an ellipse. Following this conclusion, he proposes a new form for equation (1), to ensure for the coefficient of restitution e_{out} identical values to the initial coefficient of restitution considered initially for the calculus of the coefficients of equation, namely:

$$F=Kx^{\alpha}\left[1+\frac{8(1-e)}{5e}\frac{\dot{x}}{v_0}\right]. \tag{24}$$

In equation (24) the parameters have the same meaning as the ones from equation (1). The equation proposed by Flores differs from the equation proposed in [8] only by the values of coefficients and therefore, it can be integrated by the methodology described above.

4. Results

In order to validate the proposed method, there are presented the hysteresis curves and the impact force variation, for two values of impact coefficient, and the plot of dependence between the coefficient of restitution e_{out} given by equations (1) and (24) as functions of the initial coefficient of restitution, $e_{in}\equiv e$.

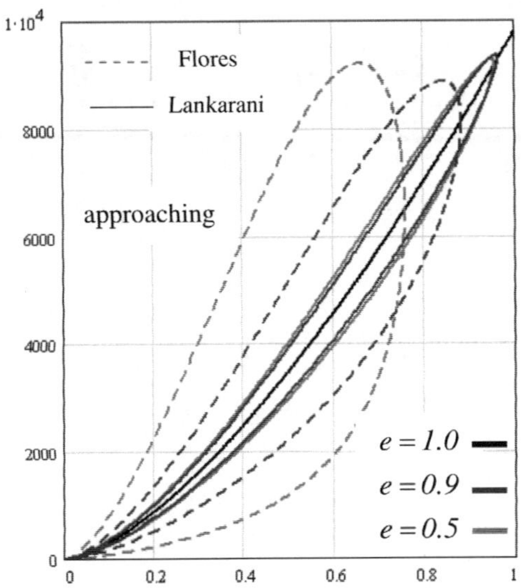

Figure5. Hysteresis loops.

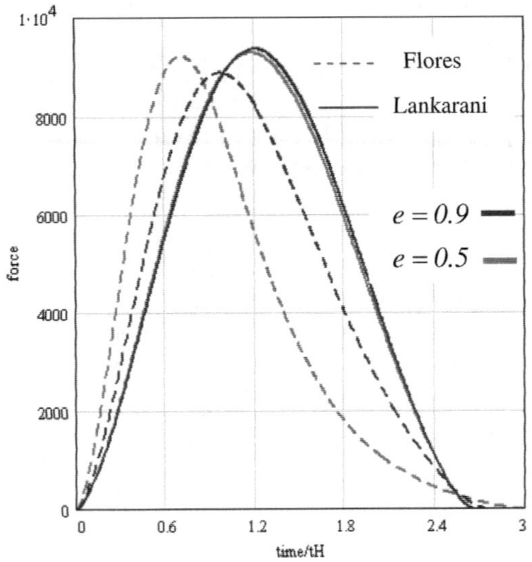

Figure 6. Impact forces versus time.

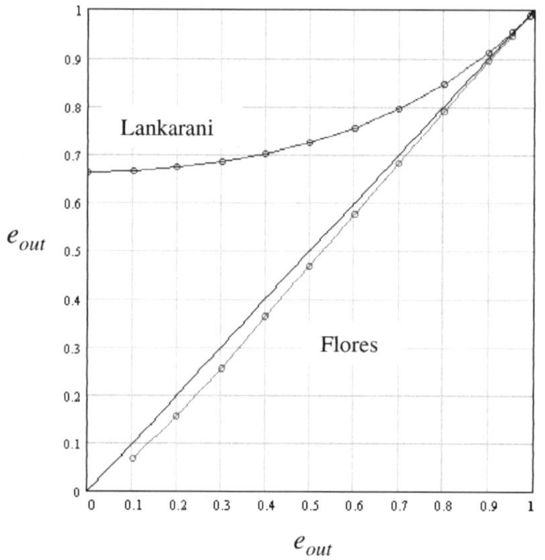

Figure 7. Dependences $e_{out} = e_{out}(e_{in})$.

The plots from figure 7 are identical to the ones presented by Flores and this confirms the exactness of the integration method for the equation.

5. Conclusions

The current paper presents a method for integration of the equation describing the dumped impact of two spheres.

The equation of the model is a second order nonlinear differential equation. The method for integration consists in two steps: first, finding by analytical calculus a first integral, allowing for the calculus of maximum indentation. The second stage concerns the numerical integration of first order nonlinear equation. The integration is made separately, as the impact process is divided in two phases: attaching or compression and detaching or expansion. The numerical procedure is based on Euler method. The precision of the results can be improved by an efficient domain partitioning for integration.

Finally, the variations with time for all collision characteristic parameters are obtained. The force variation in time is asymmetrical for approaching and detaching domains, and its maximum decreases with increasing coefficient of restitution.

The model presents a hysteresis loop, closed in the origin.

The agreement between the results obtained in the present work and the results from literature confirms the suitability of the integration method.

Acknowledgments

This paper was supported by the project "Progress and development through post-doctoral research and innovation in engineering and applied sciences– PRiDE - Contract no. POSDRU/89/1.5/S/57083", project co-funded from European Social Fund through Sectorial Operational Program Human Resources 2007-2013.

References

[1] Brach R M 1991 *Mechanical Impact Dynamics, Rigid Body Collisions* (New York: Wiley)

[2] Djerassi S 2009 *Collision with friction; Part A: Newton's hypothesis. Multibody Syst. Dyn.* **21**, 37–54

[3] Djerassi S 2009 *Collision with friction; Part B: Poisson's and stronge's hypotheses. Multibody Syst. Dyn.* **21**, 55–70
[4] Kane T R 1984 *Stanford Mechanics Alumni Club Newsletter 6*
[5] Goldsmith W 1960 *Impact, The Theory and Physical Behaviour of Colliding Solids* (Edward Arnold, Sevenoaks)
[6] Timoshenko S 1970 *Theory of Elasticity*, (Mcgraw-Hill College)
[7] Hunt K H and Crossley F R E 1975 *J. Appl. Mech.* **7**, 440
[8] Lankarani H M and Nikravesh P E 1990 *J. Mech. Des.* **112**, 369
[9] Horvay G and Veluswami M A 1980 *Hertzian Impact of Two Elastic Spheres in Presence of Surface Damping, ACTA MECANICA*, **35**. (Springer)
[10] Mathews J H and Fink K D 1999 *Numerical Method Using MATLAB* (Prentice Hall)
[11] Flores P, Machado M, Silva M T and Martins J M 2011 *Multibody System Dynamics*, **25**, 357

Journal of Physics: Conference Series **585** (2015) 012009 doi:10.1088/1742-6596/585/1/012009

Structural hysteresis model of transmitting mechanical systems

M Ruderman and T Bertram

Institute of Control Theory and Systems Engineering, TU-Dortmund, Dortmund, Germany

E-mail: mykhaylo.ruderman@tu-dortmund.de

Abstract. We present a structural hysteresis model which describes the dynamic behavior of transmitting mechanical systems with a hysteretic spring and damped bedstop element, both connected in series. From the application point view this approach can be used for predicting the transmitted mechanical force based only on the known kinematic excitation. Using the case study of an elastic gear transmission we show and identify a hysteresis response which multivariate behavior depends on an internal state of the bedstop motion.

1. Introduction

The mechanical transmission elements can exhibit strongly pronounced nonlinear characteristics, in particular in relation to the lightweight structures, complex mechanical assemblies, and frictional interfaces. A phenomenological and system-oriented modeling of such mechanical structures frequently leads back to the use of elementary and compound nonlinear operators like saturation, dead-zone, Prandtl-Ishlinskii play and stop type, Preisach, and others. Several of them are of the hysteresis type and exhibit a memory effect. For more details to the hysteresis type operators we refer to [1],[2],[3],[4]. Besides the mentioned hysteresis operators the differential models of hysteresis have gained a wide acceptance, in particular in the filed of structural mechanics. One of them, which is particularly suitable to describe the hysteresis restoring forces, is the model originally proposed by Bouc [5] and elaborated by Wen [6]. A survey on the past and recent developments and implementations of the Bouc-Wen hysteresis model is provided in [7]. For a compact analytical representation of the frictional interfaces in mechanical systems a number of physics- and empirically-motivated dynamic friction models has been proposed over the last two decades. For an overview of the development of friction models and characterization of friction force dynamics we refer to [8].

Our recent work is concerned with developing and evaluating a structural hysteresis model of transmitting mechanical systems. The model we aim to introduce is case-specific and is targeted towards the class of mechanical systems which include the damped bedstop motion and nonlinear (hysteresis) spring(s) connected in series. The proposed modeling approach aims to predict the transmitted mechanical force based only on the kinematic excitation. Often, the latter constitutes the single known value measured on the interface to the system environment.

2. Force transmitting mechanical structure with hysteresis

2.1. General consideration

A transmitting mechanical system with the damped bedstop element and nonlinear hysteresis spring is shown in Figure 1. Both mechanical transmission elements are connected in series through an internal rigid interface which degree of freedom is denoted by z. We note that a particular affiliation of the mechanical transmission elements and the total number of degrees of freedom in the system depend on the application case. However, the principal structure of a system which includes bedstop and hysteresis nonlinearities can be assumed like this. Further, we will provide a case study which matches the model structure presented in Figure 1.

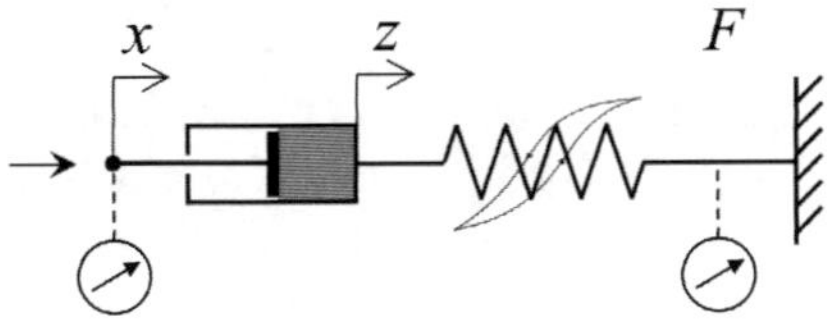

Figure 1. Mechanical structure of the force transmitting system with a bedstop motion and hysteresis spring. The kinematic excitation x and transmitted force F are the single input and output values measured in the system

The overall transmitting system undergoes a kinematic excitation in terms of a time-variant relative displacement $x(t)$ which acts on the left-hand side interface from the environment. The right-hand side interface is connected to the ground of the basic system which takes on, or more specifically consumes, the transmitted force $F(t)$. Both physical quantities constitute the system input and output and are indicated in Figure 1 as measurable. Such mechanical structures (substructures) can be found in numerous applications like vehicle and bogie suspension systems, wheel-tire systems, active dampers [9], gear mechanisms [10], and others. The principle difference to the common Maxwell model, which is represented by a purely viscous damper and elastic spring connected in series, are the nonlinear characteristics of both mechanical transmission elements. On the one hand, the damped relative motion is subject to the bedstop constraints. That is the kinematic excitation of the bedstop element transforms into the force excitation of the hysteresis spring up to the system attains a dead stop state. We note that the internal state $z(t)$ provides a transition between the kinematic and force excitation. On the one hand, the nonlinear hysteresis spring exhibits multiple equilibrium states and can dissipate energy by itself without additional dampers.

Examining in details the mechanical structure shown in Figure 1, we can infer that the composite system provides both the forwards and feedback couplings. An interplay between the dynamics of both nonlinearities occurs through the friction on a sliding interface on the one hand, and through the dead stop state of a rigid contact on the other hand.

2.2. Elastic gear transmission

As a case study we consider an elastic gear transmission which serves to boost the actuator drive torque and to provide a reduced angular motion to the output load. Such systems can be found predominantly in robotics [11] but also in machinery, aeronautic, and medical engineering. Since evermore lightweight and compliant structures are deployed in technology the torsion related issues have to be regarded when analyzing the dynamic system behavior.

The experimental setup of an elastic gear transmission is shown in Figure 2. Approaching the general mechanical structure introduced previously in Figure 1 the output shaft behind the gear transmission is locked. The free-motion torque sensor applied between the gear unit and locked-load mechanism allows to measure immediately the transmitted torque. The kinematic excitation at the free (left-hand side) end of the drive train is provided by a BLDC (brushless

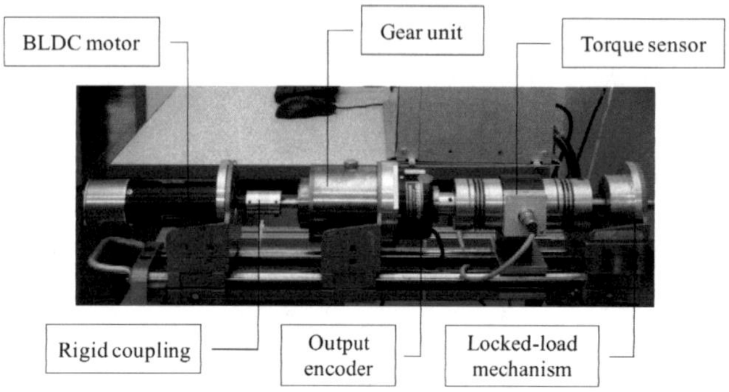

Figure 2. Experimental setup of elastic gear transmission. The output shaft behind the torque measurement is locked. The kinematic excitation is realized through the controlled angular displacement of the motor shaft.

direct current) motor through the controlled angular displacement of the motor shaft. We note that for the constrained motion experiments with the locked output shaft no measurements using the output encoder are required. The input angular displacement x and the transmitted output torque (further as force) F are the single system observations we are interested in. In between these values the complex mechanical assembly, including the couplings and gear mechanisms, constitutes a grey-box system for which the structure depicted in Figure 1 is assumed.

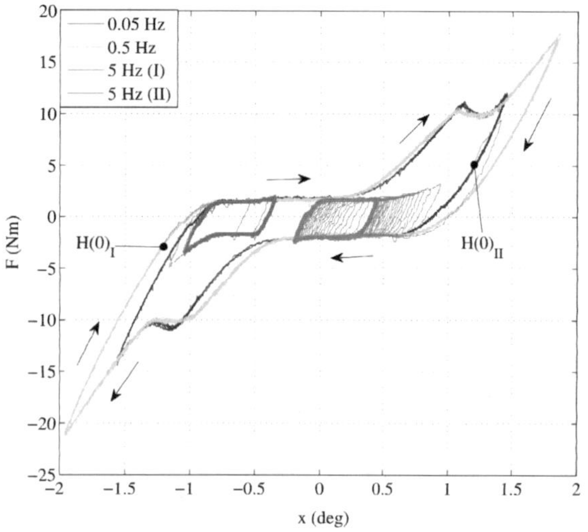

Figure 3. Measured hysteresis force response of the elastic gear transmission. The kinetic excitation of the angular displacement x is applied at frequencies 0.05 Hz, 0.5 Hz, and 5 Hz. The hysteresis response at 5 Hz is shown for two different initial states I and II. The transmitted force F is measured between the gear output and locked rotary shaft.

In Figure 3, we report the experimental results of the force response obtained at the periodic kinematic excitation of different frequencies. The measured transmitted force is shown over the angular displacement. For the sake of clarity, the angular displacement is denoted in the coordinates of the gear output, i.e. with a nominal gear reduction of 160:1. Three sinusoidal sequences of 0.05 Hz, 0.5 Hz, and 5 Hz have been applied consecutively, with a short break between each experiment. The measured hysteresis curves are symmetric to the point of origin and feature an apparent major loop obtained at 0.5 Hz excitation. The hysteresis response at 5 Hz is recorded twofold starting from two different initial states indicated in Figure 3 by $\mathrm{H}(0)_{\mathrm{I}}$ and $\mathrm{H}(0)_{\mathrm{II}}$. It can be seen that the initial trajectories of both hysteresis curves at 5 Hz excitation coincide well with the main hysteresis loops measured at lower frequencies. Afterwards, the hysteresis branching appears due to the impact of the bedstop motion. Further, we note an obviously decreasing region in the thick of the increasing hysteresis branches between ± 1.1 deg and ± 1.3 deg, and that in the first and third quadrants of hysteresis loop. It points to a certain relaxation range of the mechanical structure or to a presence of secondary backlash effects which ar not captured by the assumed bedstop motion.

3. Modeling

3.1. Bouc-Wen like hysteresis model

To represent a nonlinear mechanical spring with hysteresis the overall restoring force H has to be computed in an analytical way as a function of the relative displacement z. One of the most widely accepted models for mechanical structures is a differential model proposed by Bouc [5] and generalized by Wen [6]. This "semi-physical" model connects the relative displacement (originally deformation) and restoring force through a first-order nonlinear differential equation implying hysteresis. By choosing the appropriate values of unspecified parameters it is possible to vary substantially the shape of the hysteresis loops. The application specific tuning of the hysteresis shape parameters and numerous extensions of the Bouc-Wen hysteresis model are reported in the literature. For more details and a survey we refer to [12, 7].

In this work, the inelastic restoring force is described using a Bouc-Wen like hysteresis model

$$H(z,t) = w\, S(z)\, z(t) + (1-w)\, S(z)\, h(t)\,. \tag{1}$$

The equation (1) constitutes a weighted superposition of an elastic (z-dependent) and a hysteretic (h-dependent) contribution. The weighting coefficient $0 < w < 1$ provides the relation between a purely elastic $w = 1$ and purely hysteretic $w = 0$ force response. The hysteresis state h is given in its differential form by

$$\dot{h}(t) = \alpha\, \dot{x}(t) - \beta\, |\dot{x}(t)|\, |h(t)|^{n-1} h(t) - \gamma\, x(t)\, |h(t)|^{n}\,. \tag{2}$$

The parameters α, β, and γ constitute the basic hysteresis shape control. Note that once the control parameters β and γ are zero the state equation (2) becomes the simple elastic case which represents a linear spring. The exponential parameter $n \geq 1$ assigns the smoothness of transition from the elastic to the post-elastic branch. For the small values ($1 \leq n < 2$) the transition is smooth while for large n values the transition is abrupt. Over the years, the original Bouc-Wen model has been extended several times. Many new parameters have been added to fit a variety of hysteretic shapes [12]. However, it has been found that several parameters of Bouc-Wen like hysteresis models are functionally redundant. Even for the earliest version of the Bouc-Wen model, which is given by equation (2), a redundancy can be removed by setting $\alpha = 1$.

Another modeling issue than the modification of the Bouc-Wen hysteresis equation (2) in concerned with mapping of the stiffness characteristic curve. The original form as well as several later formulations of the Bouc-Wen like hysteresis models (see [7] for survey) imply the single stiffness coefficient instead of $S(\cdot)$ in equation (1). However, the underlying elastic stiffness characteristics can be nonlinear, for which the most common hardening effect is one of the typical phenomena. Therefore, we stress that a nonlinear stiffness map $S(z)$ can be directly integrated into a Bouc-Wen like hysteresis model as provided in equation (1). In this case the displacement-dependent stiffness map is captured by a piecewise linear function with multiple characteristic sampling points $\{z_i\}$ over the total domain $\{z_i\} \in Z \subset \mathbb{R}$. An alternative way to reshape the overall inelastic restoring force is to use an appropriate polynomial function $S(z) : z \mapsto \mathbb{R}_+$ like in [13].

3.2. Modified Maxwell-Slip representation of pre-sliding friction

On the sliding interface of the bedstop element the friction is considered mainly as a function of the relative displacement. Assuming a limited motion range of the bedstop slider no velocity- and acceleration-dependent effects of the transient friction are taken into consideration. At these simplifying conditions the mechanisms of the pre-sliding friction [14] are predominant that includes the Coulomb friction and the pre-sliding hysteresis behavior. The latter means that the friction-displacement hysteresis curves arise at the motion onset and motion reversals. The hysteresis branches saturate at a constant level of the friction force which, in the simplest case, is represented by a Coulomb friction. Note that considering a time-invariant normal load and adhesion coefficient a straightforward assumption of the constant Coulomb friction force is made.

One of the ways to describe the pre-sliding friction behavior bases on the Prandtl-Ishlinskii stop-type (PIS) operator [2, 4] which is widely used for modeling systems with hysteresis. Given the Prandtl-Ishlinskii play-type (PIP) operator [2]

$$y(t) = p[x](t)\,, \tag{3}$$

also known as mechanical backlash, the PIS operator is obtained by

$$y(t) = s[x](t) = x(t) - p[x](t)\,. \tag{4}$$

The transfer characteristics of the PIS operator are shown in Figure 4. The single threshold parameter of the PIP operator characterizes the saturation level of the PIS one. Note that using additional scaling factors for x and y the branches of the transfer characteristic can be reshaped.

In context of modeling the pres-sliding friction the PIS operator describes a simplified elastoplastic characteristic of contacting surfaces, given in force-displacement coordinates. That is the surface asperities in contact undergo first a linear elastic deformation before breaking loose. Afterwards, the sliding surfaces behave like a plastic-type element until the next motion reversal. We note that the PIS transfer characteristics are also referred to as Maxwell-slip element when modeling the dynamic friction [8]. The Maxwell-slip (elastoplastic) element is represented by a massless block placed on a rubbing surface and connected to the input through a linear spring. The block is either sticking or slipping depending on the relative displacement x applied to the opposite end of connecting spring. The stiffness coefficient of the spring and maximum (break-away) force characterize the single Maxwell-slip element and determine the horizontal lines and slopes of the transfer characteristics shown in Figure 4. Connecting in parallel multiple Maxwell-slip elements with individual stiffness and break-away properties one obtains a hysteresis map with nonlocal memory characteristics. Note that a larger number of elastoplastic elements is assumed higher accuracy of mapping the pre-sliding hysteresis behavior can be achieved. However, this can claim a higher effort for identifying parameters of the distributed model. The Maxwell-slip structure also denoted as Maxwell-slip model has been successfully applied in several dynamic friction models like Leuven and Generalized Maxwell-slip (GMS) one [8].

For the sake of convenience and in terms of keeping the number of free parameters as low as possible we apply the so called Modified Maxwell-slip (MMS) model [15] of pre-sliding friction. The MMS model implies the single elastoplastic element connected to the input through a spring, in a similar way to the elementary Maxwell-slip element. However, the connecting spring is no longer linear and exhibits a saturating elastoplastic deflection until the next motion reversal. The displacement-dependent stiffness is assumed to decrease exponentially thus providing a smooth hysteresis transition to the plastic sliding. The transfer characteristics of the MMS element are shown in Figure 5. We note that each motion reversal gives rise to a branching transfer. The MMS model in its differential form is given by

$$\dot{y}(t) = |\Omega|\,\dot{x}(t)\,K \exp\Big(-K\big|x_r(t)\big|\Big) \tag{5}$$

with

$$x_r(t) = \begin{cases} 0, & \text{when motion reverses at } t_r \\ \int\limits_{t_r}^{t} \dot{x}(t)\,dt, & \text{else}\,. \end{cases} \tag{6}$$

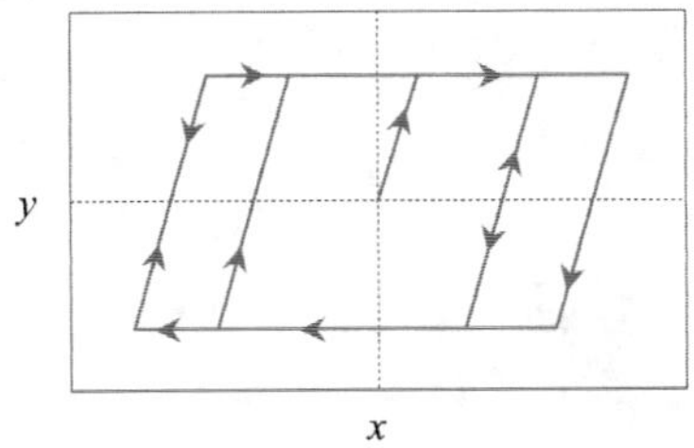

Figure 4. Prandtl-Ishlinskii stop-type (PIS) operator (in context of friction modeling also known as Maxwell-slip element). The slope transitions are reversible.

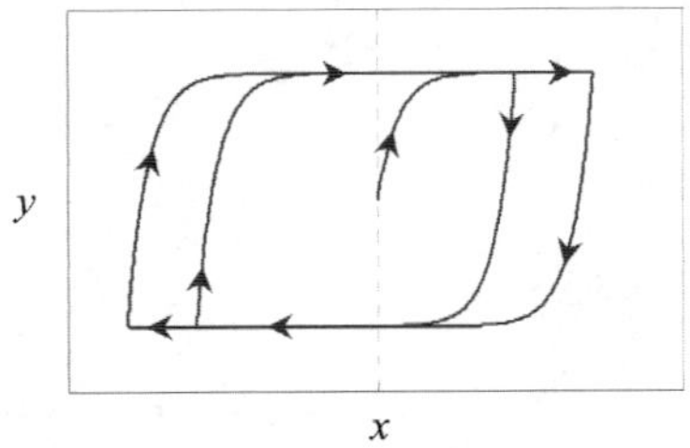

Figure 5. Modified Maxwell-slip (MMS) element for representation of pre-sliding friction. Each input reversion induces a novel (exponentially) converging curve.

The variable stiffness capacity

$$\Omega(t_r) = \mathrm{sgn}\big(\dot{x}(t)\big)\, F_c - y(t_r) \tag{7}$$

memorizes the last reversal state each time the relative motion changes the direction. The equation (7) insures that the hysteresis branch converges at the constant Coulomb friction F_c independent of the reversal state $y(t_r)$. The parameterizable initial stiffness K determines the exponential rate with which the friction force proceeds towards the plastic sliding. We note that the MMS model has a local and neither non-local memory characteristic owing to the Ω property. It means that only the last motion reversal is stored into the memory. The knowledge about previous friction states is erased once the motion direction changes. Nevertheless, the applied MMS model allows to map the smooth hysteresis trajectories in the force-displacement coordinates and to describe the pre-sliding friction using only two parameters to be identified.

3.3. Force transmission

The bedstop motion introduced in Section 2 constitutes a combination of the kinematic backlash and damped relative motion due to the friction. Since the bedstop element is connected in series with the hysteresis spring, according to Figure 1, the question is which amount of the force is transmitted to the output of overall mechanical structure. We note that due to nonlinear and constrained behavior of the bedstop element and hysteresis spring neither of classical approaches for modeling the viscoelasticity, like Maxwell model, Kelvin-Voigt model, or combination of them, can be efficiently used. Instead, the Bouc-Wen like hysteresis model and MMS model of pre-sliding friction defined so far are arranged in the manner that accomplishes the structure provided in Figure 1.

Let us consider two different modes of the executed bedstop motion as schematically shown in Figures 6 and 7. While the dead stop is not reached and the plunger moves at the applied kinematic excitation x the friction force F_f arises on the sliding interface. The same amount of the reactive force is transmitted to the output of the bedstop element and furthermore to the spring, so that $F_{out} = -F_f$ can be assumed. At the same time, the reactive force F_{out} acts upon the hysteresis spring so that the relative position $z(t)$ changes. Once the dead stop is reached (see Figure 7), i.e. $z(t) = x(t)$, the kinematic excitation acts immediately upon the hysteresis spring so that $F_{out} = H(z)$. When the direction of the kinematic excitation changes the bedstop element remains first in the dead stop until the hysteresis spring becomes fully relaxed. Afterwards, the bedstop element sets in motion and the friction force occurs in the opposite direction, similar as depicted in Figure 6. An interplay of the bedstop motion and hysteresis spring dynamics provides the overall transfer characteristics of the system.

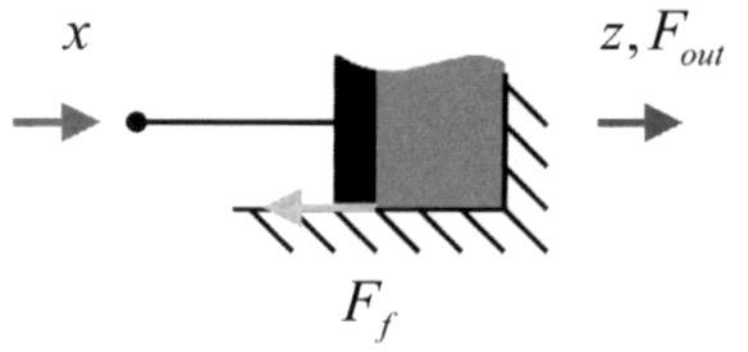

Figure 6. Schematic representation of the force transmission at bedstop motion.

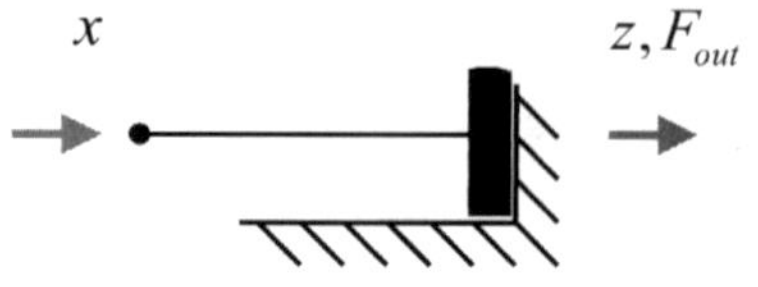

Figure 7. Schematic representation of the force transmission at dead stop.

We note that the backlash size of the system is unknown and has to be identified together with other model parameters. The backlash serves as a saturation operator of the kinematic excitation when computing the friction force on the basis of relative velocity

$$F_f(t) = \mathrm{MMS}\Big(\dot{x}(t) - \dot{z}(t)\Big) . \tag{8}$$

At the same time, the backlash appears as a dead-zone (DZ) operator [4] when considering the kinematic excitation of the hysteresis spring so that

$$F_{out}(t) = H\Big(\mathrm{DZ}[x](t), t\Big) . \tag{9}$$

The dead-zone size is the same as that of the backlash. However, the left and right dead-zone limits are time-variant and depend on the instantaneous $z(t)$ value.

4. Identification of model parameters

The derived model of transmitting mechanical system provides several degrees of freedom concerning the bedstop motion dynamics and shape of the hysteresis spring. A full decomposition of the overall system response in the associated terms appears as not directly feasible, due to an internal feedback and the lack of internal state measurements. Therefore, an appropriate system excitation aims to reduce as possible the impact of one of the transfer elements while inducing a characteristic behavior of the other. The system response obtained in this way can be considered as an approximated decomposition of both nonlinearities.

The identification of free model parameters is accomplished in two stages. As a first step, the parameters of the Bouc-Wen like hysteresis model are identified using the experimental data of the main hysteresis loop measured at 0.5 Hz excitation (see Figure 3).

The experimental data of an increasing hysteresis branch shown in Figure 8 is used to determine the stiffness characteristic curve $S(\cdot)$ described by a piecewise linear function. A compact set of overall twelve sampling points $\{x_i, F_i\}$, depicted in Figure 8 over the measurement, is selected with respect to the curvature of the hysteresis branch. Following, the stiffness characteristic curve is determined in due consideration of the force bias F_1 related to the hysteresis loop so that

$$\Big\{|z_i| = x_i, F_i - F_1\Big\} \in S(z) . \tag{10}$$

Note that a symmetric hysteresis spring is assumed so that an absolute z value is included in equation (10). The $S(z)$ values between the sampling points are obtained using the linear

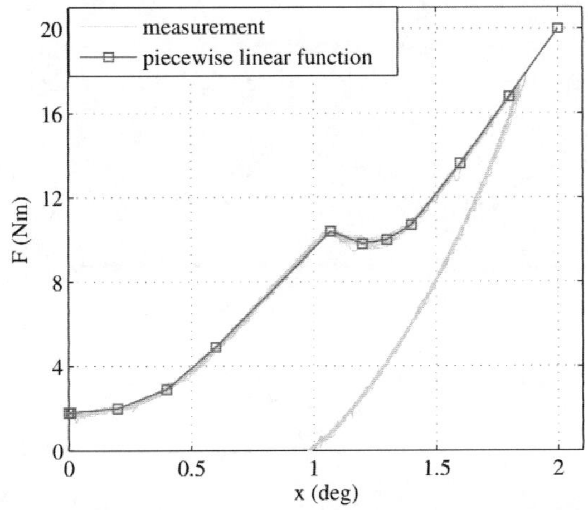

Figure 8. Piecewise linear function of the stiffness characteristic curve depicted over the measurement. The experimental data are taken from the major hysteresis loop measured at 0.5 Hz sinusoidal excitation. The determined stiffness characteristic curve is illustrated together with the corresponding sampling points.

interpolation. We note that the major hysteresis loop measured at a relatively slow smooth excitation does not include the internal branching attributed to the bedstop motion. Under these conditions the hysteresis spring is assumed to be solely contributing to the output force.

In order to determine the residual parameters of the Bouc-Wen like hysteresis model the experimental time series of multiple hysteresis cycles obtained at the same 0.5 Hz excitation are used. Applying the standard Levenberg-Marquardt nonlinear fit algorithm the parameters are identified in a least-squares sense. In Figure 9, the overall response of the modeled hysteresis spring is shown over the measurement. It is worth noting that since the hysteresis produced by the Bouc-Wen like model is rate-independent the singe frequency excitation is sufficient to determine the full set of the model parameters.

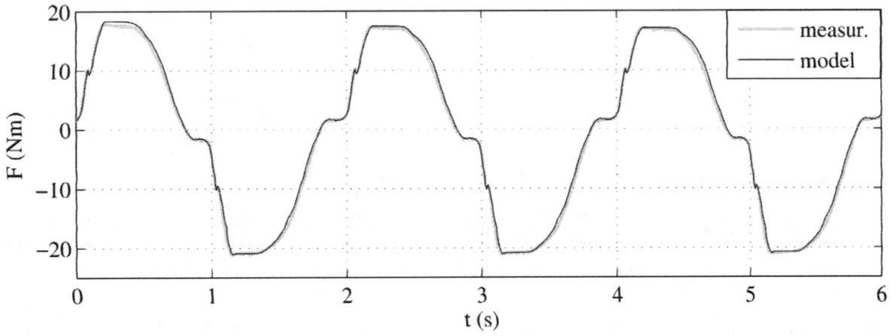

Figure 9. Identified and measured transmitted torque as time series. The parameters of the Bouc-Wen like hysteresis model are determined from the main displacement-force hysteresis loop measured at 0.5 Hz sinusoidal excitation.

The parameters of the modeled bedstop motion, among of the MMS friction model and the backlash size, are determined from the experiment with a variable excitation frequency. For this purpose, a linear down-chirp excitation is applied which covers the frequency range between 5 Hz and 0.1 Hz. We note that since neither frictional conditions nor backlash size of the system are known in advance a continuous variation of the excitation frequency aims to disclose the interaction between the bedstop element and hysteresis spring. The measured force response is shown in Figure 10 as a time series together with the response of the overall identified model. The residual model parameters are identified from the data in a least-squares sense.

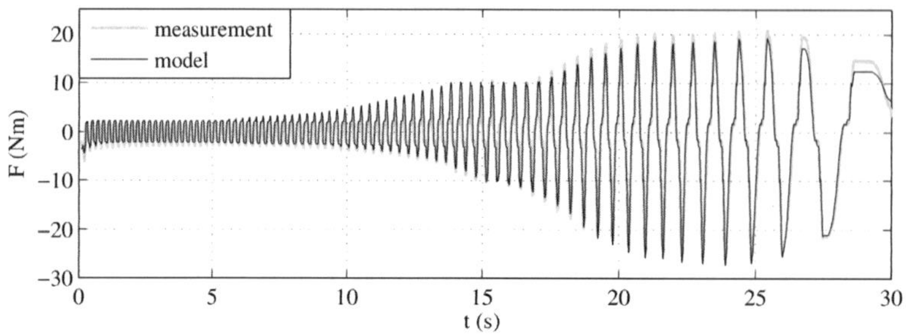

Figure 10. Identified and measured torque response with hysteresis spring and bedstop motion. The linear down-chirp excitation with a bandwidth 5–0.1 Hz is applied.

It can be seen that the impact of the pre-sliding friction is predominant at times below 10 seconds. At this stage the hysteresis spring is marginally stressed so that no significant restoring force is contributing to the overall system output. In particular, we can infer that the hysteresis spring response becomes perceptible once the applied frequency is low enough to drive the system predominantly in a dead stop state (see Figure 7). Further, it is worth noting that no knowledge of the initial hysteresis state, which is $h(0)$ in terms of the equations (1) and (2), contributes to the model uncertainties which are visible in Figure 10 at times between 10 and 15 seconds.

5. Conclusions

We have proposed and evaluated in experiments a novel structural hysteresis model of transmitting mechanical systems. The problem of predicting the mechanical force transmitted through a structure with the damped bedstop motion and nonlinear hysteresis spring has been addressed. We have shown that the computation of the output force based only on the known kinematic excitation can be realized in an analytical form. It has been shown that both system nonlinearities are interacting with each other in a forward and feedback manner. It has also been emphasized that an internal not-measurable state gives rise to a multivariate hysteresis behavior observable at the system output. We suggest that the identification of the model parameters has to be carried out using such system excitations which decompose as possible the impact of both nonlinearities. The results of this paper can be of use for an accurate modeling and simulation of transmitting mechanical structures in various physics and engineering fields.

References

[1] Mayergoyz I D 1986 *Physical Review Letters* **56**
[2] Krasnosel'skii M and Pokrovskii A 1989 *Systems with Hysteresis* (Berlin: Springer)
[3] Krejci P 1991 *Applications of Mathematics* **36**
[4] Brokate M and Sprekels J 1996 *Hysteresis and Phase Transitions* (Berlin: Springer)
[5] Bouc R 1967 *Proc. 4th Conference on Nonlinear Oscillations* (Prague)
[6] Wen Y K 1976 *ASCE Journal of the Engineering Mechanics Division* **102** 249
[7] Ismail M, Ikhouane F and Rodellar J 2009 *Archives of Computational Methods in Engineering* **16** 161
[8] Al-Bender F and Swevers J 2008 *IEEE Control Systems Magazine* **28** 64
[9] Choi S B, Lee S K and Park Y P 2001 *Journal of Sound and vibration* **245** 375
[10] Seyfferth W, Maghzal A J and Angeles J 1995 *Proc. IEEE Conf. on Robotics and Autom.* (Nagoya) p 3027
[11] Ruderman M, Hoffmann F and Bertram T 2009 *IEEE Transactions on Industrial Electronics* **56** 3840
[12] Ma F, Zhang H, Bockstedte A, Foliente G C and Paevere P 2004 *Journal of Applied Mechanics* **71** 342
[13] Ruderman M, Aranovskiy S, Bobtsov A and Bertram T 2012 *Automation and Remote Control* (*Preprint* .)
[14] Al-Bender F, Lampaert V and Swevers J 2004 *Chaos* **14** 446
[15] Ruderman M and Bertram T 2011 *Proc. 18th IFAC World Congress* (Milano) p 10764

Experimental Highlight of Hysteresis Phenomenon in Rolling Contact

S Alaci[1], D A Cerlincă[1] , F C Ciornei[1], C Filote[2] and G Frunză[1]

[1] Mechanics and Technology Department, University of Suceava, 13 Universitatii Street, 720229 Suceava, Romania
[2] Faculty of Electrical Engineerig and Computer Science, University of Suceava, 13 Universitatii Street, 720229 Suceava, Romania

E-mail: stalaci@yahoo.com

Abstract. In literature, the hysteresis phenomenon in rolling contacts is studied considering both rolling friction and sliding friction. Removal of sliding friction in experimental tests from a concentrated contact is a serious challenge. The paper proposes a method and presents a device ensuring pure rolling between two identical discs, normally loaded. Using photoelastic material for the two rolling discs, by means of photoelastic method, the hysteresis phenomenon due to rolling friction is qualitatively confirmed.

1. Theoretical aspects

Mechanics of rigid body theory is based on the hypothesis of undeformable bodies that is mathematically expressed by the fact that the distance between any two points of the body does not change in time, regardless of the magnitude of applied loads. However, when explaining the rolling friction phenomenon, this assumption must be overlooked. Voinea, [1], considers a disc of radius r, resting on a rigid horizontal plane and loaded by a horizontal force F and its own weight G, as in figure 1.

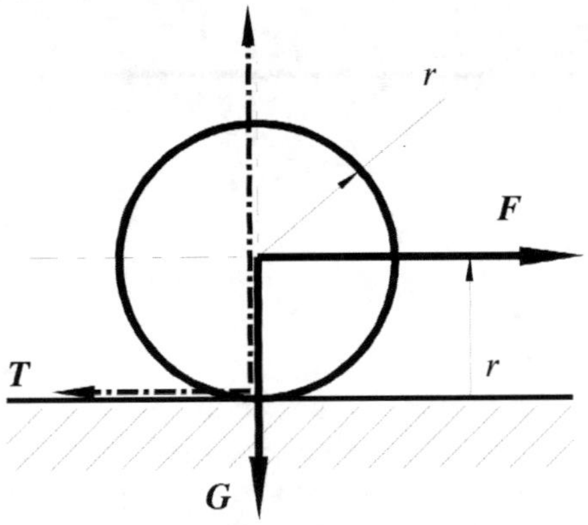

Figure 1. Contact between a rigid cylinder and a rigid half-space.

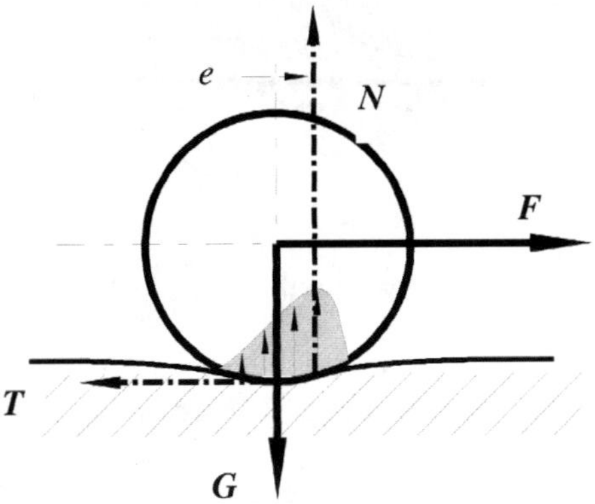

Figure 2. Contact between an elastic cylinder and an elastic half-space.

The reaction force from the plane has two components: the normal reaction, $\boldsymbol{N}$ and the tangential reaction $\boldsymbol{T}$ (or the friction force). The equilibrium equation for moments written with respect to the contact point leads to:

$$F = 0 \tag{1}$$

According to equation (1), but contrary to practice, the body could be brought to rolling motion by any small force. In order to explain the plane resistance to rolling, the hypothesis of rigid plane must be disregarded and therefore, a deformable plane should be considered. Under this assumption, the contact between the cylinder and the plane doesn't occur along a generatrix but on a surface on which a pressure distribution arises, asymmetrical with respect to the vertical line passing through the center of the cylinder, as seen in figure 2. The pressure distribution ought to be asymmetrical as the force torsor reduced to the initial contact point should contain the normal force N and the moment M_r with respect to the initial contact point. The point of reduced force torsor can be chosen to be situated at a distance e from the theoretical contact point where the moment is zero, the condition being:

$$M_r < eN \tag{2}$$

The maximum value of eccentricity e for which the disc still rests immobile is named rolling coefficient and is denoted by s. Since $e \le s$, by multiplying by N this inequality, one obtains:

$$M_r = e\,N \tag{3}$$

As long as the next condition is fulfilled, the body will not effectuate rolling motion:

$$M_r = eN \le sN, \tag{4}$$

The maximum value of eccentricity e is named coefficient of rolling friction.

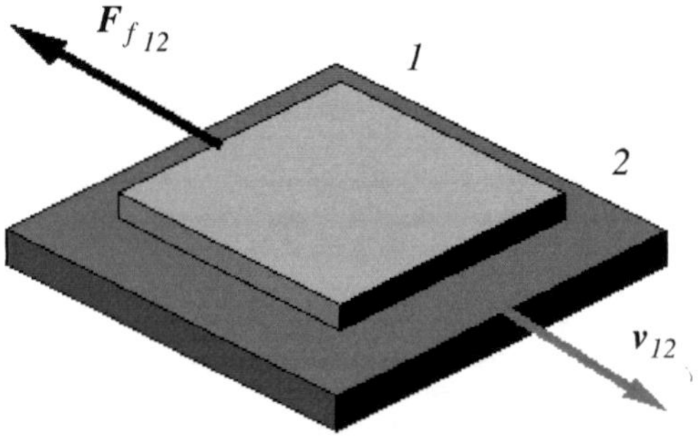

Figure 3. Case of pure sliding occurring between two bodies.

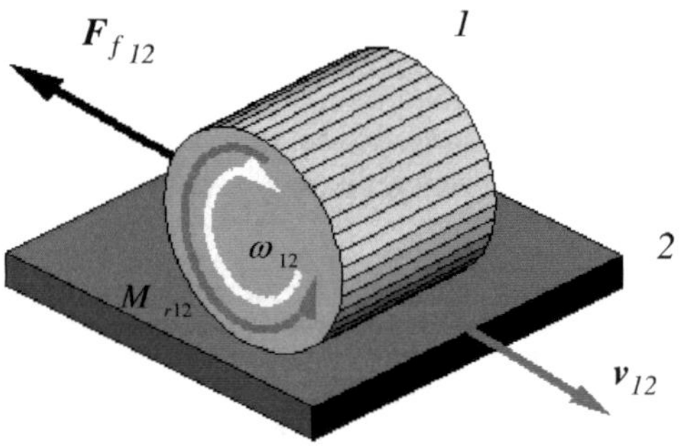

Figure 4. Case of general plane motion when both sliding and rolling friction exist.

There are practical situations when between two bodies only relative translation motion occurs, as represented in figure 3, and just the sliding friction as friction type. For most of the cases, [2,3,4,5], the relative motion between two bodies is a general spatial motion. For the situation of a contacting cylinder 1 and a prismatic part 2, the upper body 1 is in relative plane motion with respect to the body 2 and this relative motion is characterized by the velocity $\boldsymbol{v}_{12}$. In figure 4, the relative motion between the bodies is characterized by the relative angular velocity vector ω_{12}, parallel to the axis of the cylinder and by the relative velocity between the theoretical points of contact, v_{12}. At first sight, if the following condition is fulfilled:

$$v_{12} = \omega_{12} R \tag{5}$$

and, for a value of sliding friction coefficient large enough (where R is the radius of the cylinder), it could appear that the pure rolling condition is satisfied. Actually, due to deformability, [6], the instantaneous axis of relative rotation doesn't coincide with the initial contact generatrix and therefore the pure rolling condition is violated, [7].

2. Constructive requirements

According to the above observation, pure rolling will occur between the two bodies only in the case in which the instantaneous axis of relative motion between the two bodies coincides to the initial contact straight line and this conducts to the idea of using two deformable cylinders of precise equal radii, of similar materials, [6], pressed one against the other. The contact surface is a plane one and the axis of cylinders are symmetrically placed about it for any load. The geometrical and loading symmetry ensures the plane shape of contact surface and will contain the axode of relative motion. Pure rolling occurrence requires, besides the geometrical condition of equal cylinders radii, precise equality of angular velocities of cylinders.

3. Photoelasticity considerations

The photoelastic method was chosen to study the stress field produced between the two discs. The main advantages of the method consist in the option of global stress field analysis from the tested parts and in the fact that the stress field structure remains unaltered as the probes are not supposed to direct contact. The main disadvantage of the method is the limitation of applying the method only for plane models. For spatial models, a special study method – freezing stresses method, must be applied, [8]. The basis for the photoelastic studies is the birefringence phenomenon. Since the proposed testing device scheme assumes a plane model loaded by a plane force system, the usual photoelastic methodology is applied.

For the case of plane stress state in a body, this stress state will be completely characterized in a Cartesian coordinate system by a second order tensor, ***T***:

$$\boldsymbol{T} = \begin{bmatrix} \sigma_x & \tau_{xy} \\ \tau_{xy} & \sigma_y \end{bmatrix} \tag{6}$$

where σ_x, σ_y are the normal stresses and $\tau_{xy} = \tau_{yx}$ are the shear stresses. The eigenvalues of this tensor are the values of principal normal stresses:

$$\lambda_1 = \sigma_{max}, \ \lambda_2 = \sigma_{min} \tag{7}$$

The principal directions of the tensor corresponding to the principal normal stresses are characterized by the orthogonal versors $\boldsymbol{e}_1$ and $\boldsymbol{e}_2$, and between the versors $\boldsymbol{i}$ and $\boldsymbol{e}_1$ the angle φ appear, figure 5.

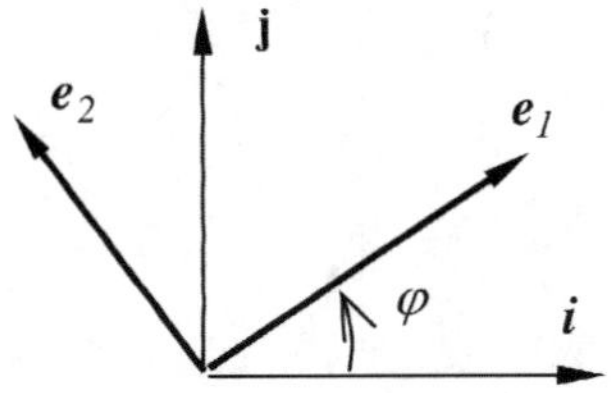

Figure 5. Cartesian coordinate system and the directions of principal stresses.

Three families of curves are used in photoelastic studies:

Journal of Physics: Conference Series **585** (2015) 012010 doi:10.1088/1742-6596/585/1/012010

Isochromatics, defined as geometrical locus of points for which the maximum shear stress is the same

$$\tau_{max} = (\sigma_{max} - \sigma_{min})/2 = const. \tag{8}$$

Isoclinics, defined as the locus of the points in the specimen along which the principal stresses are in the same direction.

$$\varphi = atan[2\tau_{xy}/(\sigma_x - \sigma_y)] = constant. \tag{9}$$

Isoclinics and isochromatics are overlapped and cannot be separated.

The isopachs are the curves along which the sum of normal principal stresses is constant or the mean normal stress is constant.

$$\sigma_1 + \sigma_2 = \sigma_x + \sigma_y = tr(\boldsymbol{T}) = const. \tag{10}$$

where *tr*(***T***) represents the *trace* of stress tensor ***T***.

For a model subjected to a plane stress state, the complete stress tensor in a point is found by knowing the characteristic parameters of the three curves passing through the considered point.

From the three families of curves, the most used are the isoclinics. In white light, the isoclinics appear as a succession of colored curves. In polarized light, these curves become more clear, the image of isochromatics field is a succession of dark and bright curves in white and black. As the load of a region increases, the isochromatics gather tighter and the brightness of the region increases. For the next considerations, is important the observation that an unloaded contour appears dark. In the case of loaded contour, it becomes bright due only to principal normal hoop stress.

4. Design principles for the device

As previously précised, for the study of rolling friction effect, between the two cylinders, rolling motion without sliding must exist. The presence of sliding friction alters the symmetry of isochromatics field. In the isochromatics field, the presence of a concentrated force on the contour is evidenced by the occurrence of a pattern named „peacock eye", [9], figure 6. The level curves are circles tangent in the loading point.

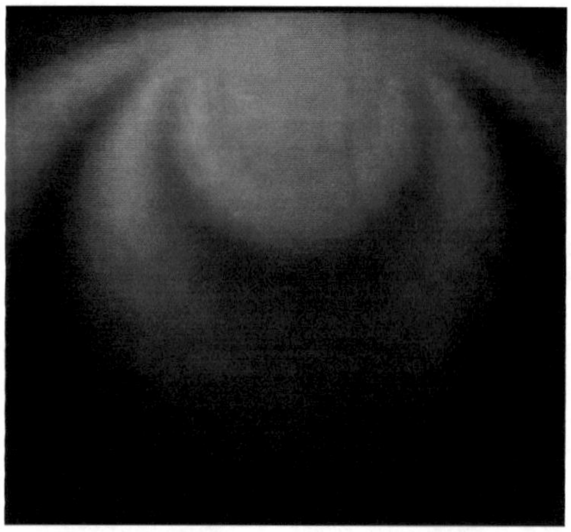

Figure 6. Isochromatic field near a concentrated normal load on boundary.

In order to emphasize the effect of friction force upon the isochromatics field, a dry friction contact was modeled using finite element method , for friction coeficient $\mu = 0$ and $\mu = 0.05$ respectively. The results are presented in figure 7.

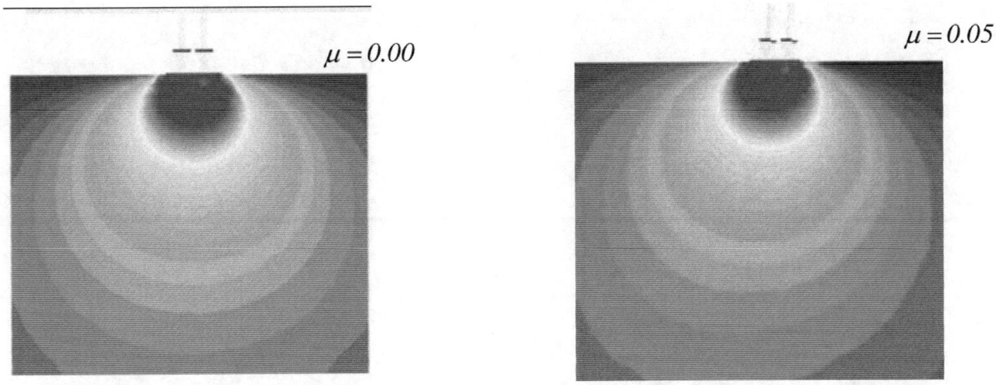

Figure7. The isochromatics field from an elastic half-plane normally loaded, without friction (left) and with friction (right).

The presence of friction leads to asymmetrical loading and therefore to the occurrence of asymmetry in isochromatics field.

With the above considerations, the following device scheme is proposed, Fig. 8. Two identical discs, with parallel axis, made of photoelastic material, are brought into contact and rotated in opposite ways with strictly equal angular velocities. The prefect equality of the angular velocities of the wheels is accomplished by mounting them on two rotating shafts, the kinematical linkage between the shafts consisting in a gear train. The contact region between the cylinders must be free to allow the polarized light waves to pass.

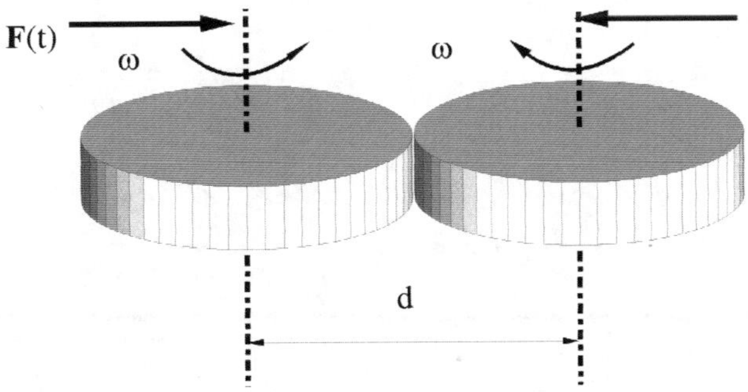

Figure 8. Rolling contact between two identical cylinders.

To this end, the gears from the shafts must have smaller diameters compared to the photoelastic discs and the shafts must be actuated by two gears ensuring opposite rotations direction of the discs, figure 9.

In order to load the discs radial, the distance between the gears 1 and 4 must be adjustable. The condition of equality for angular velocities of gears 1 and 4 plus the possibility of adaptable axis distance lead to the conclusion that one of the gears 1 or 4 must be a sun gear. The driving motion is made using a d.c. motor, because it allows adjusting the angular velocity via the voltage. The reduced available space on the polariscope table for attaching the testing device imposed the use of a fifth gear on which shaft a belt wheel was placed.

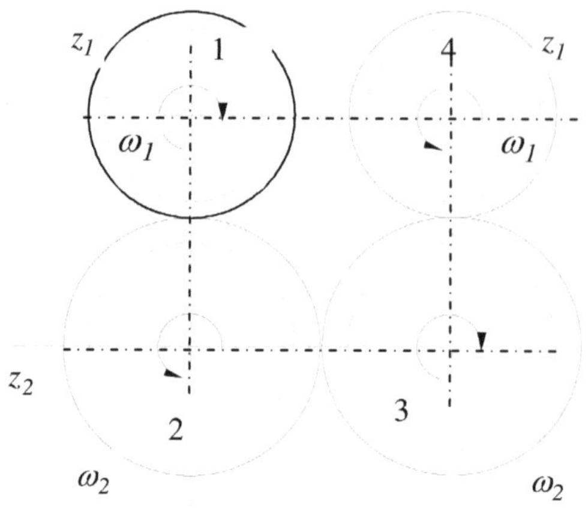

Figure 9

The mechanism from figure 9 , ensures the conditions that the rotations of the shafts 1 and the one on which the photoelestic discs will be fitted should be rigorously equal, due to equal number of teeth of the gears 2 and 3, and 1 and 4, respectively. The inconvenience of this mechanism is that, due to fixed positions of shafts, cannot ensure radial loading of photoelastic discs. To respond to this requirement, the final version of the mechanism was proposed, as presented in figure 10. Thus, the photoelastic disc united with gear 1 has a fixed center while the second disc has the center on the lever P of a differential mechanism.

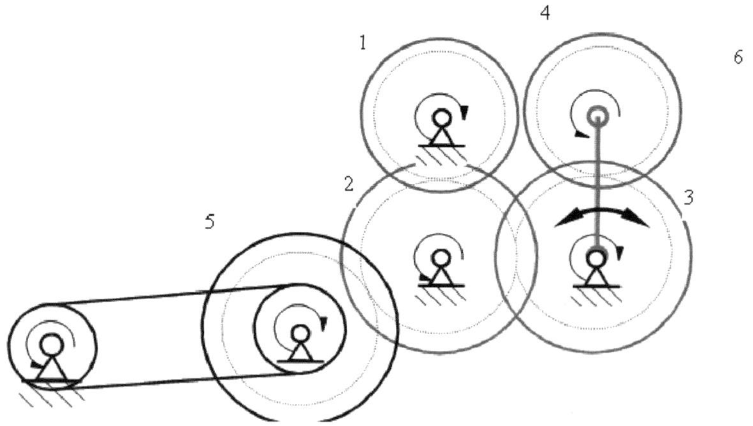

Figure 10. Final solution for driving into rotation the photoelastic discs.

As long as the discs are not in contact, the disc 4 presents its own rotation around the center ω_4, that is composing with the velocity of differential and planet wheel carrier according to Willis formula, [10].

$$\frac{\omega_4 - \omega_p}{\omega_3 - \omega_p} = \frac{z_4}{z_3} \tag{11}$$

When the contact between the photoelastic discs happens, the planet wheel carrier becomes immobile, $\omega_p = 0$, the relation 10 becomes:

$$\frac{\omega_4}{\omega_3} = \frac{z_4}{z_3} = \frac{z_1}{z_2} = \frac{\omega_1}{\omega_2}. \tag{12}$$

In addition, the angular velocities ω_2 and ω_2 are equal in magnitude but of opposite directions and therefore, finally $\omega_4 = -\omega_1$. The experimental testing device is presented in Figure 11, made based on the kinematical scheme from Figure 10. The driving motion is made using a d.c. motor and a belt transmission.

Figure 11. Detail and general view of the experimental device.

5. Experimental results

The experimental device was attached to the stage of a VISHAY P500 Series Polariscope. The rotative speed is measured using a non contact tachometer. The experimental test aimed observing the isochromatics field obtained for different loads and angular velocities. For very small loads, one can notice that the contact between the two discs appears faintly lighted due to first isocromatics occurrence. During increasing load period, the next isochromatics appear, situated in the next vicinity of contact point. With further increased load, the lighted region dimensions increase and the distinction between isochromatics is more difficult to be made due to brightness of the loaded region. For a strong load rise, the brightness increases further but the angular velocity of the discs decreases considerable due to augmented internal friction from the material of the two discs.

In figure 12 the experimental isochromatics of the two discs are shown, correlated to the rotating directions. The most interesting aspect appears in figure 12 b), where in the region of contact exit from the contours of the discs, a brightness gradually decreasing can be observed. This brightness can be caused by:

- Plastic deformations;
- Friction forces presence;
- Hysteresis phenomenon.

Within the assumption of plastic deformations, the entire periphery of the discs should be lighted on the entire contour.

The hypothesis that sliding friction occurs between the two discs could be caused by different values of the diameters of the two photoelastic discs and possible errors from the condition of equal angular velocities of the shafts of the photoelastic discs, due to execution errors and device fitting errors. If the assumption of friction is considered, than in the contact point the friction forces are equal in magnitude but with opposite directions and the isochromatics fields from the two discs should be anti-symmetrical.

The only feasible remaining hypothesis is the one considering that the brightness is produced by the hysteresis phenomenon. The material from the vicinity of contact point enters the contact and is rapidly deformed. After leaving the contact, the material remains loaded for a while and this fact determines the occurrence of brightness on the discs contours in the contact exit region.

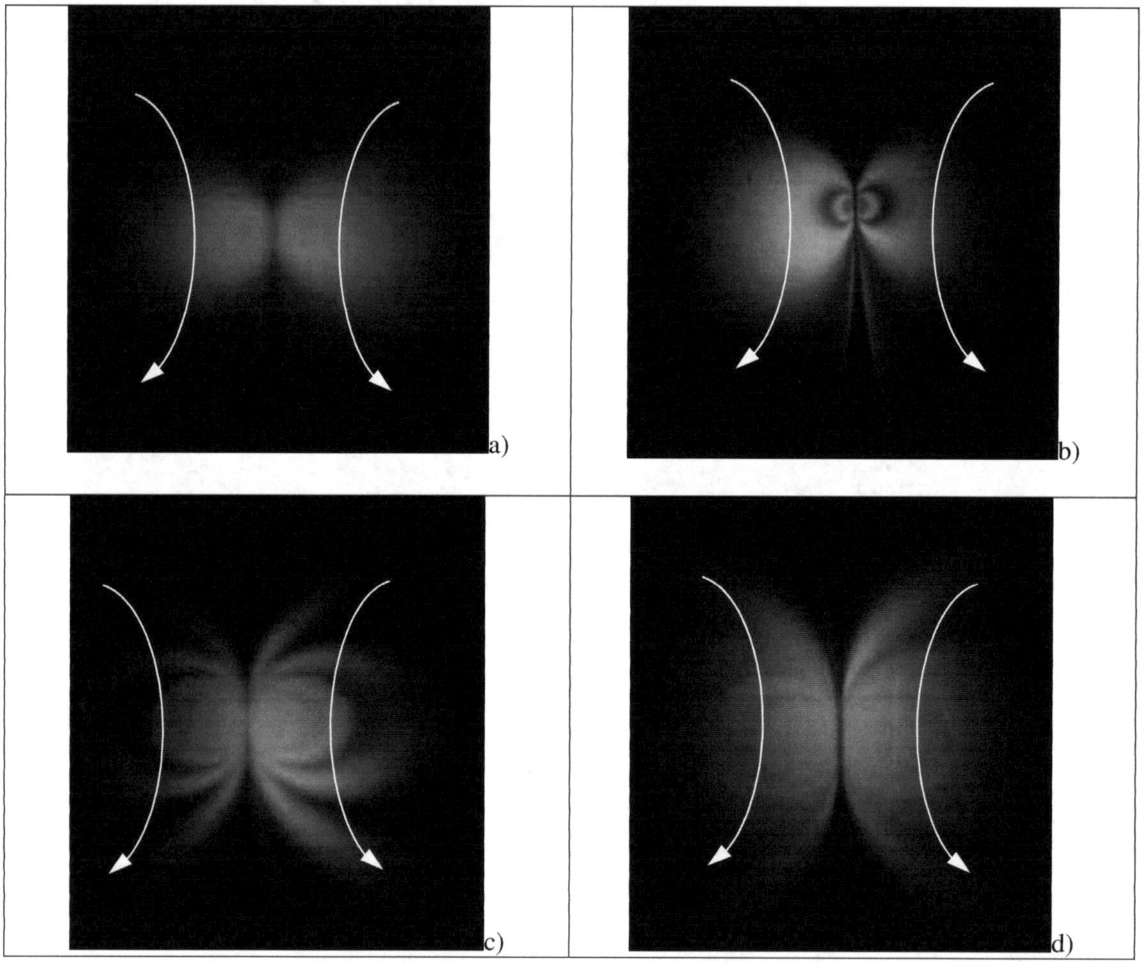

Figure 12. Experimental isochromatics; increasing loads from a) to d).

For increased values of load F, the isochromatics field is symmetrical in the absence of motion. Under rolling conditions, for increased normal loads, the hysteresis phenomenon cannot be marked because with increasing load, the intensity and extent of brightened area increases too. This is due to the fact that the stresses (and their optical effect, the brightness) due to hysteresis become insignificant compared to the stress field produced by the Hertzian contact.

6. Conclusions

The paper presents a device designed for the study of rolling contacts based on the photoelastic method. Most of the photoelastic researches are made on probes in static conditions.

The test rig shows a silent running, confirmed by the isochromatics shape from the two discs as during running at a rotating speed of 2400 rpm the aspect is quasi static.

The contact hysteresis phenomenon was revealed.

During experimental tests, it was noticed that for increasing load, the angular velocity of the motor drive decreases. In order to maintain a constant rotating speed, a new motor drive is required, having greater power or a stiffer characteristic.

Another solution for obtaining uniform motion is mounting a disc flywheel on one of the shafts.

The device can also be used for the study of rolling friction with sliding, by modifying the discs, having different radii but with fixed distance between axes.

For further studies, a proposed improvement of the testing device consists in using photoelastic discs of equal diameters but employing pairs of exchangeable wheels used in accomplishing a controlled sliding.

Acknowledgement:
This paper was supported by the project "Progress and development through post-doctoral research and innovation in engineering and applied sciences – PRiDE - Contract no. POSDRU/89/1.5/S/57083", project co-funded from European Social Fund through Sectorial Operational Program Human Resources 2007-2013.

References
[1] Voinea R, Ceausu D and Voiculescu V 1983 *Mecanica* (EDP Bucuresti)
[2] Bulanov E A 2007 *Journal of Machinery Manufacture and Reliability* **36**, No. 5, 430–433 © Allerton Press, Inc., 2007. Original Russian Text © E.A. Bulanov 2007, published in *Problemy Mashinostroeniya i Nadezhnosti Mashin* **5** 42–46
[3] Tabor D 1955 *Proc. R. Soc. Lond.* A **229** 198-220
[4] Greenwood J A, Minshall H and Tabor D 1961 *Proc. R. Soc. Lond.* A **259** 480-507
[5] Pinegin S.V. 1976 *Trenie kacheniya v mashinakh i priborakh* (Resistance to Rolling in Machines and Devices) (Moscow: Mashinostroenie)
[6] Johnson K L 1987 *Contact Mechanics* (Cambridge U P)
[7] Huang L A 2011 *Concise Introduction to Mechanics of Rigid Bodies: Multidisciplinary Engineering* (Springer)
[8] Coker E G , Filon L N G and Jessop H T 1957 *A Treatise on Photo-Elasticity* (Cambridge U.P.)
[9] Frocht M M 1949 *Photoelasticity* I, II (John Wiley & Sons Inc).
[10] Uicker J, Pennock G and Shigley J 2010 *Theory of Machines and Mechanisms* (Oxford U P)

Analysis of thermally stimulated processes in new phenanthroline derivatives suitable for optoelectronic devices

Marius Prelipceanu[1], Lucian Cojocariu[1], Adrian Graur[1], Sigurd Schrader[2]
[1]Faculty of Electrical and Computer Engineering, Stefan cel Mare University, Suceava 720229, Romania
[2]Department of Physics Engineering, University of Applied Sciences Wildau, Germany

E-mail: mprelipceanu@eed.usv.ro

Abstract. Thermally stimulated processes have been studied in thin films of phenanthroline derives to describe the states that had been localized. Ultraviolet Photoelectron Spectroscopy had checked out before thin films of new pyrrolo[1,2-a][1,10] phenanthroline derivatives [1], in order to further applications in optoelectronic devices. The investigated compounds have an electronic band structure which is due to the substituent (R=NO2, Cl) induced transformations of molecular orbitals. Thermally stimulated discharge currents (TSDC) attested dipolar and charge transport mechanisms. Each TSDC peak has been assigned to elementary processes that are different utilizing in addition dielectric spectroscopy (DES), and the mean trap depths have been approximated from thermally stimulated luminescence (TSL) curves.

1. Introduction

The candidates that are promising for the technology of solid-state devices, organic light emitting diodes (OLEDs) [1] in particular, are the Pyrrolo[1,2-a][1,10]phenanthroline (RA) derivatives. Due to their ionization potential and electron affinity, the above-mentioned materials are captivating for the optoelectronic devices, allowing electron injection at the cathode [2,3]. Because of the specific chemical structures of these materials, molecular orbitals can be tuned by inserting acceptor or donor groups, obtaining emissive or electron transporting/hole blocking materials [4].

The mix between the hole transporting materials like polyparaphenylenevinylene (PPV), and pyrrolo[1,2-a][1,10]phenanthroline as electron transporting in hetero-layer structures is a technique to develop new performances of OLEDs in contrast with single layer diodes. The requirement of finding and describing new materials suitable as charge transporting layers in OLEDs is critical from this perspective. Electronic processes in phenanthroline derivatives have been studied, and then, the localized states [5-7] have been examined by utilizing the thermally stimulated techniques. The basic valence electronic structure was given by the UPS measurements, while the electronic transport processes in bulk materials [8] have been described by the thermally stimulated techniques (TSDC, TSL) and impedance spectroscopy (DES). For a better quality of these devices, the attention needs to be distributed to materials synthesis and to the study of the electronic structures of these devices which allows apprehending of their conduct in organic electronic devices [9]. On the other hand, the continuous decrease in fabrication size leads to a significant increase in the effects of thermally induced / assisted phenomena which should be carefully addressed [10].

The chemical structure of pyrrolo[1,2-a][1,10]phenanthroline derivatives is presented in Fig. 1. [11-14]. By spin coating, the thin films were placed in clean conditions and onto flat silicon. The thickness of the film determined from the ellipsometric measurements was 100 nm for the study samples. We utilized a toroidal grating monochromator (TGM2-beam line) covering the photon energy range from 5 to 190 eV. At the room temperature [1], the photoelectrons were gathered with an ADES 400 angle resolving spectrometer system. During the experiment, performed at BESSY (Berlin) [1]., the base pressure was 2 x 10-10 mbar.

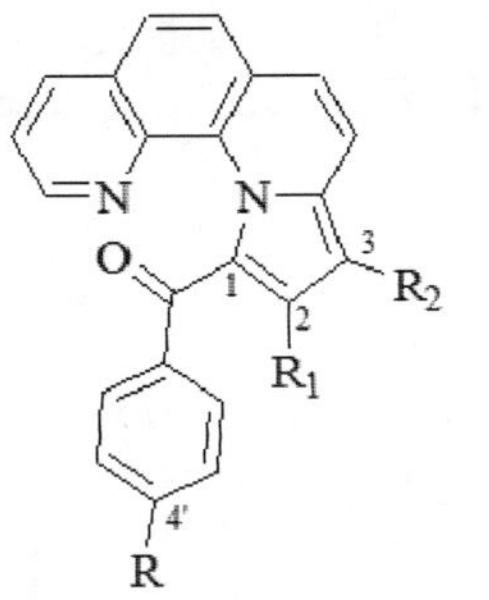

RA1: R= NO_2, R_1=R_2= $COOCH_3$
RA2: R= Cl, R_1=R_2= $COOCH_3$

Figure 1. Chemical structure of pyrrolo[1,2-a][1,10] phenanthroline derivatives investigated

The photoelectron spectra were measured at an incidence angle of 450, in normal emission, for incident energies 50 eV. By utilizing solution-casting technique, thermally stimulated analysis samples were prepared: polymer films were cast on the surface of freshly evaporated aluminum copper (RA) from a chloroform solution, allowing the solvent to evaporate slowly. We heated the films up to 360 k for 10 hours, and then, to eliminate any residual solvent, dried them in high vacuum (5 x 10-6 Torr) for 2 hour at Tg, gained using a constant heating rate of 0.5 K/s. All measurements were made with a custom configuration [2]. With a temperature range between 75 and 450 K, we performed the TSL measurements. We used a 1000 W Xenon lamp as an excitation source and, at a constant temperature, we illuminated the samples for 15 min, and, after we switched off the light, we heated them with a linear rate of 0.1 K/s. Evaporating a second aluminum electrode (150 nm) under UHV conditions [2], we utilized a sandwich configuration in order to achieve TSDC and DES measurements.

2. Results and discussions

UPS spectra of compounds RA1(a) and RA2(b) is being shown by Figs. 2. Also, the same figures show the theoretical simulations for different incident photon energies of both compounds. The vacuum level (E0) has been set at zero [15], and all energy values were considered in respect with it. The incident photon energy used for the measurements presented in Fig. 2, was 50eV. For clarity, we used vertical offsets. Between experimental findings and theoretical results, a good agreement had been found, and, through this, each UPS peak has related to the specific contributions of distinct groups of occupied molecular orbitals [15-18]. A good agreement between experimental findings and theoretical results was found, allowing relate each UPS peak to the specific contributions of distinct groups of occupied molecular orbitals [15-18]. Due to the highest occupied molecular orbital (HOMO level), the lowest binding energy peak is essentially for both compounds. Also, the π-orbitals produce this emission. For compound RA1, the first peak has its onset at 2.11eV, in case of 50eV incident photon energy. For compound RA2, its onset is at 5.60eV, in case of 50eV incident photon energy. The mixture between π- and σ- states The second peak is already due to a mixture of π- and σ- states. The other peaks are related to two and three centers σ-states. As it can be seen in Fig 2, The HOMO's energy positions is As it is presented in Fig. 2, the HOMO's energy positions and of very close states do not practically change in examined compounds; therefore we can assume that there is not any shift with the molecular size [17, 18].

We will now take a closer look at the obtained spectra. When we increase the exciting energy from 40eV to 50eV, the beam electron density is strongly intensified for both compounds, due to the interaction between molecular electrons and photons for deeper orbitals that has an increased probability. In the case of compound RA1(R=NO2), at 50 eV - excitation electrons from the orbitals previously excited at 40 eV are also emitted, but the latter are occupied by electrons emitted from the deepest orbital. In this way, at 50eV-excitation the levels by 2.90 eV and 9.66 eV are poorly evidenced. The highest localized orbital is the localized orbital by 2.90 eV. This localized orbital forms at the absorption band's long wavelength limit. A rather wide band with maximum at 7.36 eV can be observed because of the bands' overlapping at 5.00 eV, 9.66 eV and 6.60 eV, and because of the electronic background from the deepest orbital (by 23.31 eV), at 50 eV incident energy. For compound RA2 (R=Cl), the 1.85 eV, 3.80 eV and 5.55 eV levels are not observed due to the same reason. This threshold is seen to displace toward small energies because of overlap with the intense band from 5.55 eV. Further present at 8.72 eV and 10.72 eV are the characteristic bands of Cl. The weak band located at 12.90 eV is further present as well. Due to the deepest orbital, the increase of current density is over 15eV. The increase can be observed at exciting energies. We improved TSD, TSL and DES measurements to describe the localized states. TSDC spectra of RA1 and RA2 are shown in Fig. 3. Spectra of RA1 and RA2 show two peaks. Both can have a dipolar origin or both can return from charge transport processes.

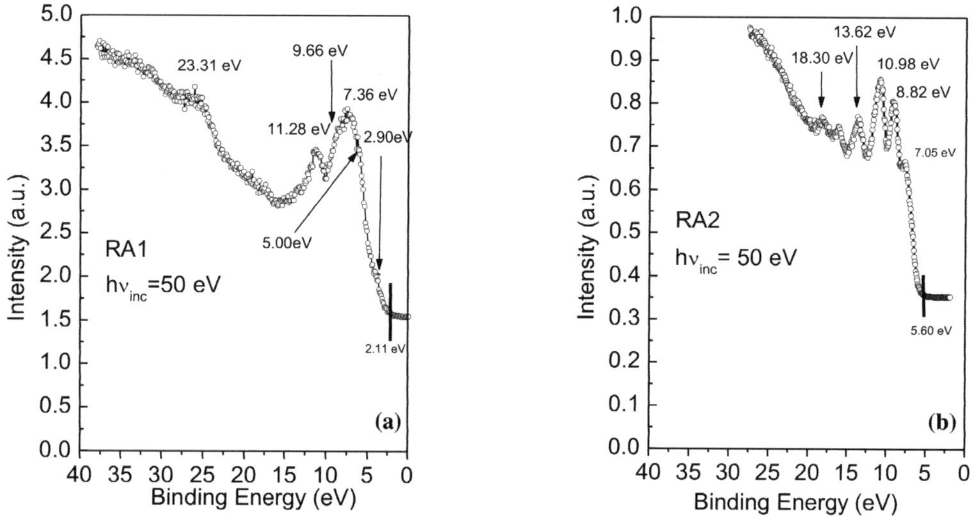

Figure 2. UPS valence band spectra of compounds RA1 (a) and RA2 (b) at 50 eV incident photon energy.

The DES measurements were carried out on both types of RA to inspect their nature. In figure 4, we can see exemplified the DES spectra for peaks II and I in RA2. As it is shown, in peak I is no evidence of any evidence of dipolar relaxation, which is very clear instead for peak II. On one hand, the scanned temperature range was established by presuming a single elementary relaxation process, and taking into account the connection between the relaxation frequency [20] and the TSD current. The single elementary relaxation process can be found described by an Arrhenius law. On the other hand, the scanned temperature range was established by using the Debye description [21-23] of the dielectric constant complex component [18, 22, 23].

Both of the peaks have different nature. While peak II has dipolar origin, peak I may be due to charge transport processes, and particularly to trap states in the material. The TSL measurements have been consequently performed to investigate the presence of the trap distributions. In Fig. 5, the resultant TSL curves for RA1 and RA2 are shown. According to Halperin and Braner [21], and based on the mean activation energy, Ea, utilizing Lushchik's formula [21-23], the symmetry factor μ_g has been estimated. In Figure 5 it can be observed the obtained values. Two trap states have been showed by all materials: for RA1 and RA 2, a shallow one at 0.14 eV, and for RA2, a deeper state at at 0.24

eV. We cannot determine if the carriers are holes, electrons or both, based only on these TSL results, but there is a chance to estimate the trap densities [23].

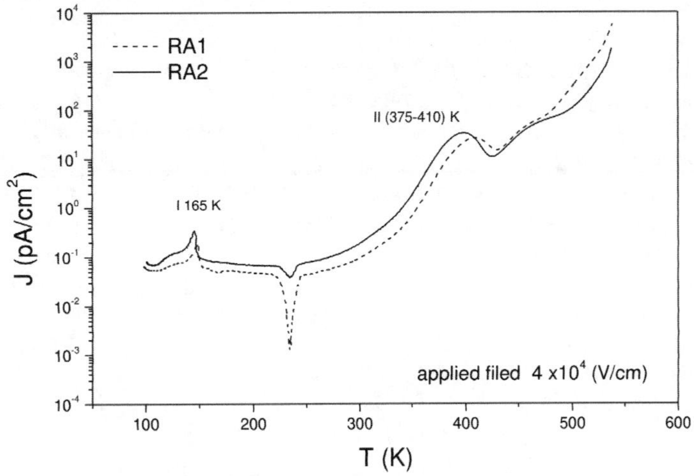

Figure 3. TSDC spectra of RA1 and RA2

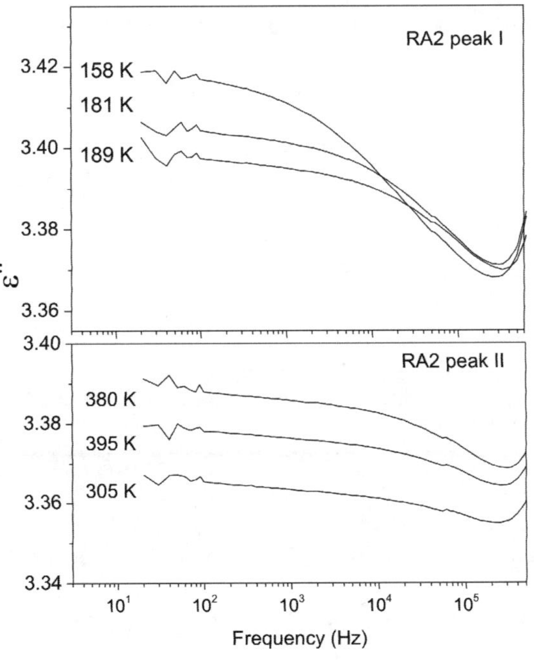

Figure 4. DES spectra peak I and peak II for RA2

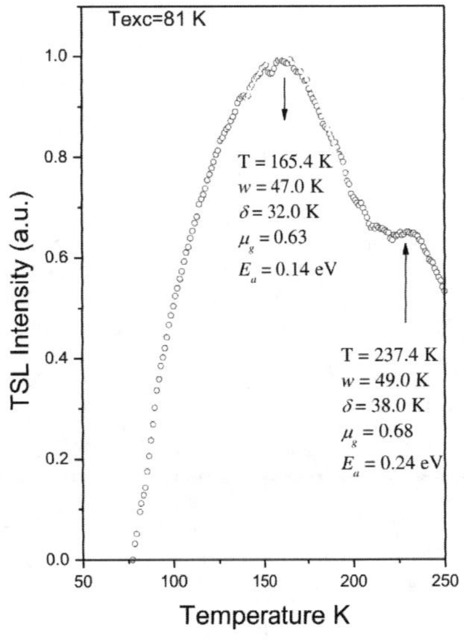

Figure 5. TSL spectra for RA1 and RA2

3. Conclusion

The UPS measurements were carried out in two different types of pyrrolo[1,2-a][1,10]phenanthroline derivatives, to induce their valence band structure, and their respective potentials of ionization. We obtained a good agreement between the experimental curves and the quantum chemical simulations of UPS spectra. This agreement suggested that the calculated dipole moments are reliable. To describe the localized states, the TSDC, DES and TSL measurements were carried out that present the existence of dipolar relaxation and charge transport processes because of the trap states. The trap states that are large enough are the deeper trap states, and they can contribute to the carrier transport in OLEDs.

Acknowledgement
The project "Q-DOC - Improving the quality of doctoral studies in engineering sciences to support the development of the knowledge society – Contract POSDRU/107/1.5/S/78534" and Romanian National Program II, Young Research Team, contract no. 107/06.08.2010 have partially supported this work. The "Q-DOC - Improving the quality of doctoral studies in engineering sciences to support the development of the knowledge society – Contract POSDRU/107/1.5/S/78534" was co-funded from European Social Fund through Sectorial Operational Program Human Resources 2007–201.
We gratefully acknowledge Dr. Paolo Imperia from Australian Nuclear Science and Technology Organization for helpful measurement, Prof. Dr. Liviu Leonte and Dr. R. Danac, for providing us the phenanthroline derivatives.

References

[1] Prelipceanu M, Prelipceanu O S, Leontie L, Danac R, 2007 Physics Letters A **368** 331.
[2] Imperia P, Casu M B, Schrader S, Falk B, Jandke M, Strohriegl P, 2001 Synthetic Metals **124** 83.
[3] Weaver M S, Lidzey D G, Fisher T A, Pate M A, Brien D O, Bleyer A, Tajbakhsh A, Bradley D, Skolnick M S, Hill G, 1996 Thin Solid Films **273** 39.
[4] Leontie L, Druta I, Danac R, Prelipceanu M, Rusu G I, 2005 Progress in Organic Coatings, **54**, 3.
[5] Prelipceanu M, Leontie L, Danac R, Prelipceanu O S, 2006 Romanian Conference on Advanced Materials, Bucharest-Magurele, Romania, Proc.
[6] Bettenhausen J, Greczmiel M, Jandke M, Strohriegl S, 1997, Synth. Met. **91** 223.
[7] Zugravescu I, Petrovanu, M, 1987 *3+2 Dipolar Cycloaddition*, Rom. Acad. Publ. House, Bucharest (in Romanian).
[8] Druta I, Andrei M, Aburel P, 1998 Tetrahedron **54** 2107.
[9] Druta I, Dinica R, Bacu E, Humelnicu I, 1998 Tetrahedron **54** 10811.
[10] Dimian M, Andrei P, *Noise-driven phenomena in hysteretic systems*, Springer, New York, 2014.
[11] Dinica R, Druta I, Pettinari C, 2000 Synlett. **7** 1013.
[12] Danac R, Rotaru A, Drochioiu G, Druta I, 2003 J. Heterocyclic Chem. **40** 283.
[13] Druta I, Danac R, Barbieru R, Tapu D, Andrei M, Sci. Ann. Al. I. Cuza Univ. Iasi, S. Chem.
[14] Grem G, Leditzky G, Ullrich B, Leising G, 1992 Adv. Mat. **4** 36.
[15] Casu M, Imperia P, Schrader S, Schulz B, Jandke M, Strohriegl P, 2001 Synth. Met. **121** 1397.
[16] Imperia P, Schader S, Casu M B, Jandke M, Strohriege P, 2000 Nonlinear Optics **34** 456.
[17] Schrader S, Imperia P, Koch N, Leising G, Falk B, 1999 Proceeding of SPIE **3797** 209.
[18] Van Turnhout J, in G.M. Sessler, 1980, *Electrets, Topics in Applied Physics*, ed. (Springer-Verlag, Berlin).
[19] Vanderschueren J, Gasiot J, in P. Bräunlich, 1979 *Thermally Stimulated Relaxation in Solids, Topics in Applied Physics*, ed.(Springer-Verlag, Berlin).
[20] Halperin A, Braner A, 1960 Phys. Rev. **117** 408.
[21] Lushchik C B, 1956 Sov. Phys.-J ETP **3** 390.
[22] Opanowicz A, 2007, J. Phys. D: Appl. Phys. **40** 4980.
[23] Okachi T, Nagase T, Kobayashi T, Naito H, 2008 Thin Solid Films **517** 1327.
[24] Okachi T, Nagase T, Kobayashi T, Naito H, 2009, Applied Physics Letters, **94** 4.

6th International Workshop on Multi-Rate Processes and Hysteresis (MURPHYS2012) IOP Publishing
Journal of Physics: Conference Series **585** (2015) 012012 doi:10.1088/1742-6596/585/1/012012

Switching behaviour of two-phenotype bacteria in varying environment

G Friedman[1], P Gurevich[2], S McCarthy[3] and D Rachinskii[3,4]

[1]Department of Electrical & Computer Engineering, Drexel University, Philadelphia PA, USA
[2]Institute of Mathematics I, Free University of Berlin, Arnimallee 2-6, 14195 Berlin, Germany
[3]Department of Applied Mathematics, University College Cork, Cork, Ireland
[4]Department of Mathematical Sciences, The University of Texas at Dallas, Richardson, USA

E-mail: stephen.p.mccarthy@mars.ucc.ie

Abstract. An increasing interest in multi-phenotype species has stimulated both experimental and mathematical research. One example is bacteria which have two phenotypes and can make transitions from one phenotype to the other in response to variations in environmental conditions. We model a population of such bacteria subjected to a stochastic environmental input, which fluctuates between two conditions preferred by the phenotypes. Our interest in this model is how the average growth rate of the total population is affected by alterations to the environmental thresholds at which the transitions between phenotypes are allowed. Under certain conditions, we find that the bacteria achieve a maximum growth rate by adjusting their behavior to act in a similar manner to a non-ideal relay. In this scenario, memory helps to increase fitness. We then extend the model to include multiple competing species with different thresholds and examine the limit of distribution of population among these species and phenotypes as time increases. For this purpose, we formulate a reaction-diffusion model which involves non-ideal relays describing the evolution of the state of different species; and, a Preisach operator with time-dependent density function to account for the integral effect of the species on the environment. Formation of patterns and multiple stationary limits are shown numerically in the multi-species model.

1. Introduction

Long-term memory underpins genetic inheritance mechanisms and functioning of the immune system in higher organisms. Many regulatory networks exhibit complex dynamics and multi-stable states, also linked to memory. There is mounting evidence that microorganisms such as bacteria which have experienced different environmental histories may respond differently to current conditions and that an environmental memory can grant fitness to bacteria in the evolutionary game. For example, history dependent behaviour was shown experimentally, and quantified in the information theory framework, in *B. subtilis* [1]. Environmental memory has been shown in DNA methylation in chemotaxing bacteria [2], genetic and epigenetic phase variation mechanisms in pathogenic bacteria [3,4] and switch-type bistable systems in regulatory networks [5–8]. Further work focuses on designing biological memory switches such as a heritable switch with multiple states which are encoded into the DNA sequence [9].

Simple bacterial systems are an attractive test bed for developing understanding of adaptation mechanisms which help organisms to survive and improve fitness in a fluctuating environment. One example is bacteria capable of switching behaviours, phenotype, or state in response to fluctuations of external conditions if different phenotypes are well adapted to different conditions. In particular, *E. coli* generate multistability of gene expression states under different perturbations [10–13].

A number of conceptual, technically simple, mathematical models have been proposed to understand how certain switching strategies can help bacteria to increase fitness, which is often measured by the net population growth rate. The hybrid linear differential model proposed in [14] shows that a dynamically heterogeneous bacterial population can sometimes achieve a higher growth rate than a homogeneous one provided that the rate of a transition between phenotypes is comparable to, or lower than, the rate of variations of the environment. In this case, bacteria can anticipate sudden fluctuations of the environment by having a subpopulation ready in an appropriate phenotype before the environment changes to a state favouring this phenotype. This effect is more pronounced for predictable environments, such as periodic, but was also shown to be present for stochastic environments in narrower parameter ranges [14]. The optimal heterogeneous distribution between phenotypes is achieved dynamically by allowing some positive rate of transitions from the currently most favoured phenotype to an unfavoured one.

If the rate of transitions between phenotypes is higher than the rate of environment variations, then, under the assumptions of the model studied in [14], the maximal fitness is achieved by the responsive switching strategy whereby all the bacteria switch to the currently most favoured phenotype. However, if the responsive strategy is penalised, then it can be not optimal. The penalty can be associated with the cost of sensing the environment as bacteria have to maintain some sensory machinery to respond to changes. Another linear hybrid differential model, where the cost of sensing was implemented as an explicit reduction in the growth rates, shows that stochastic switching can confer more fitness to the population than the responsive switching in slowly varying environments [15]. This result, as well as the results in [14] for faster environments, support the idea that diversity (heterogeneity) can help improve fitness in varying conditions, that is the view very well established in ecology.

Memory can also grant fitness to bacteria. For instance, past phosphate limitation was shown to lead to a faster response to successive periods of phosphate limitation in *E. coli*, and this faster response was suggested to be survival enhancing [16]. An advantage of using strategies with memory can be shown in the framework of the game theory [17–19] applied to models where bacteria are considered as players in an evolutionary game [7, 15, 20].

Differential models proposed in [14, 15] are memoryless in slowly varying environmental conditions. In the adiabatic limit when the ratio of environment variation rate and the switching rate between phenotypes tends to zero, the state of the system is completely determined by the state of the environment. In this paper, using a modification of the model proposed in [14], we show that a long-term bistability type memory can increase fitness of bacteria in a stochastically varying environment. Our model includes a cost of switching in the form of a temporary inhibition of the reproductive activity in bacteria undergoing a transition to a different phenotype. We show that the memory confers fitness when the rates of the environment variation and switching are comparable. In this state, the system achieves dynamic heterogeneity, which is consistent with the results in [14]. However, a static memory in our model persists in the limit of slow environments. The memory is characterised by a responsiveness parameter α; the value $\alpha = 0$ corresponds to a memoryless system.

This paper has two parts. After characterising the type of persistent memory which ensures the maximal net growth rate of a population of bacteria in the first part of the paper, we discuss a model where populations with different α compete for nutrients.

More specifically, in the first part, we consider a linear model of a growing population of bacteria, which respond to variations of external conditions (environment) by switching between two phenotypes. We demonstrate a possibility of a scenario where memory can help achieve better fitness. The effect of external conditions is modelled by varying the availability of nutrients and, by this, the growth rate of each phenotype; fitness is measured by the average growth rate of the total population. It is assumed that each phenotype consumes a different type of food, for example, one consumes lactose, the other consumes glucose. A mixture of the two nutrients is supplied at a constant rate, while the proportion of the two ingredients in the mixture, measured by a variable E, varies in time. Bacteria sense changes in E and thus, by changing to the phenotype for which more food is available, can potentially increase the growth rate of the population. In this way, the system is similar to the model studied in [14]. However, instead of the binary environment, we use a continuous input $E(t)$ and assume the threshold type response of bacteria. Namely, bacteria decide to switch their phenotype after E passes a certain threshold value. Importantly, the threshold value E_1 for switching from phenotype one to phenotype two can be different from the threshold value E_2 for inverse switching. The equality $E_1 = E_2$ corresponds to memoryless bacteria, while the case $E_1 \neq E_2$ indicates the presence of memory as a bacterium has to know (remember) its phenotype in order to decide whether to switch or not at a certain value of E. This is the type of memory of a non-ideal relay (bistable switch) [21].

Some *in vitro* experimental studies give evidence that the process of changing phenotype can be stressful for bacteria. In particular, bacteria may not reproduce within a period of time preceding, during, or following this process.[1] If this is the case, a delay in growth appears to be a natural cost associated with phenotype switching. Assuming such a cost in our model, we will show that having memory can be an advantage for bacteria living in a stochastically varying environment. The memory, by introducing the difference between the thresholds, prevents switching when E varies in the interval $E_1 < E < E_2$ around the value $E_T = 0.5$ with equally available nutrients. This somewhat conservative strategy may increase the population's growth rate as small random fluctuations of E near the point E_T can keep more sensitive (responsive) bacteria, such as memoryless bacteria with $E_1 = E_2 = E_T$, or bacteria with $E_1 \approx E_2 \approx E_T$, in a transition non-reproductive state forever, thus stopping the growth. We will describe the optimal strategy of bacteria as a probabilistic binary switch with memory. At the same time, it can be shown that in a deterministic environment with a predictable pattern of variation (for example, a periodic environment) the most responsive memoryless switching strategy can be optimal.

In the second part of the paper, we allow for variability of the switching thresholds in a population of bacteria. The thresholds in each species are assumed to be symmetric with respect to $E = E_T$, that is $E_1 = E_T - \alpha$, $E_2 = E_T + \alpha$, with α varying in different species. Having shown in the first part of the paper that there is an optimal $\alpha = \alpha_{opt}$ which maximises the growth rate, and that α_{opt} depends on the parameters of the system and food dynamics, we ask the question whether species with α close to α_{opt} will outgrow the others if they compete for food. That is, we are interested in how fitness and competition act to select a certain distribution of switching thresholds in the population.

For simplicity, we assume that switching is fast-the time of a transition from one phenotype state to the other is short compared to the characteristic time of the variation of environment. Hence, a population of bacteria with a given α is modelled by the classical deterministic non-ideal relay. For one model of competing populations with α ranging over an interval of values, we show the formation of patterns resulting from the evolution. Our simulations indicate that, depending on initial data, the solution can converge to different patterns. That is, there is no single winner of the competition, or a single limit distribution. The attractor seems to be a connected continual set of stationary distributions.

[1] In these experiments, a colony of bacteria grown in a Petri dish with one nutrient is swapped to a Petri dish with another nutrient. After a period of inactivity, or a shock, following the swap, bacteria start a transition to the other phenotype which is better fit for consuming the new type of food.

Journal of Physics: Conference Series **585** (2015) 012012 doi:10.1088/1742-6596/585/1/012012

Our model is a system of reaction-diffusion equations containing discontinuous nonlinearities, the non-ideal relays, and a continuous integral of those. This integral can be interpreted as the Preisach operator [21] with a time dependent density (the density is a component of the solution describing the varying distribution of bacteria).

The first and the second parts of the paper constitute the two following sections. The paper ends with conclusions, including some discussion of the results and possible modifications and extensions of the models.

2. Model of a single population of two phenotype bacteria

2.1. Model formulation

The following model of two phenotype bacteria living in discretely switching binary environment was presented in [14]:

$$\tilde{x}' = \gamma_1\tilde{x} - \kappa_1\tilde{x} + \kappa_2\tilde{y}, \tag{1a}$$

$$\tilde{y}' = \gamma_2\tilde{y} - \kappa_2\tilde{y} + \kappa_1\tilde{x}. \tag{1b}$$

Here the function $\tilde{x}(t)$ is the population of bacteria in one of the phenotypes and $\tilde{y}(t)$ is the population in the other phenotype. The parameters γ_1, γ_2, κ_1 and κ_2 change whenever the environment changes state; γ_1 and γ_2 denote the growth rates of the phenotypes $\tilde{x}$ and $\tilde{y}$, respectively, in the current environment state; κ_1 and κ_2 are the transition rates between the two phenotypes.

In this work, we consider the case where the environment changes continuously within an interval of values, as opposed to discretely. We introduce the continuous variable E to denote which of the environmental states the system is currently in. The value of E is a measure of the relative abundance of one saturated environmental state over the other saturated state in the current intermediate (mixed) state. For example, if the environmental conditions that we consider are a glucose/lactose mix, then E would tell us how much glucose there was compared to lactose. We now make some definitions about the value of the variable E; we take the case of $E \leq 0$ to be the case where the phenotype $\tilde{x}$ is fully favoured, that is its growth rate is at its maximum and the growth rate of $\tilde{y}$ is at its minimum. Similarly, for $E \geq 1$ we have the case where the phenotype $\tilde{y}$ is fully favoured. Intermediate values of E give rise to partial favouring of one phenotype over the other. We define the value $E = 0.5$ as the environmental threshold, E_T, the point at which both phenotypes are equally favoured.

In case of discrete (binary) environment, as in [14], it is realistic to assume that the growth rates of the two phenotypes are discrete too and switch each time the environment switches. As we assume gradual changes of the environment between its continuous states, and there is only a partial favouring of one phenotype over the other, it seems natural to use functions that change the growth rate gradually as the environment fluctuates. With this in mind, we replace γ_1 and γ_2 in model (1) with functions Γ_1 and Γ_2 given by

$$\Gamma_1(E) = \begin{cases} \gamma_{unfit} + \sigma, & E \leq 0, \\ \gamma_{unfit} + \sigma(1 - E), & 0 < E < 1, \\ \gamma_{unfit}, & E \geq 1, \end{cases} \tag{2a}$$

$$\Gamma_2(E) = \begin{cases} \gamma_{unfit}, & E \leq 0, \\ \gamma_{unfit} + \sigma E, & 0 < E < 1, \\ \gamma_{unfit} + \sigma, & E \geq 1, \end{cases} \tag{2b}$$

where γ_{unfit} is the growth rate when the environment is in the state that is unfavoured by the phenotype; σ is the benefit for the environment being in the state that is preferred by the phenotype, see Figures 1a and 1b. We assume a positive transition rate κ from a phenotype unfavoured by the environment to the favoured phenotype; and, the zero transition rate from the favoured phenotype. The transition rate changes when E reaches a threshold value. That is, we assume that the parameters κ_1 and κ_2 in (1) are step functions of E,

$$\kappa_1(E) = \begin{cases} 0, & E \leq E_T + \alpha, \\ \kappa, & E > E_T + \alpha, \end{cases} \tag{3a}$$

$$\kappa_2(E) = \begin{cases} \kappa, & E \leq E_T - \alpha, \\ 0, & E > E_T - \alpha, \end{cases} \tag{3b}$$

see Figures 1c and 1d. Note that if $\alpha > 0$, then no transitions occur as long as $E_T - \alpha < E(t) < E_T + \alpha$.

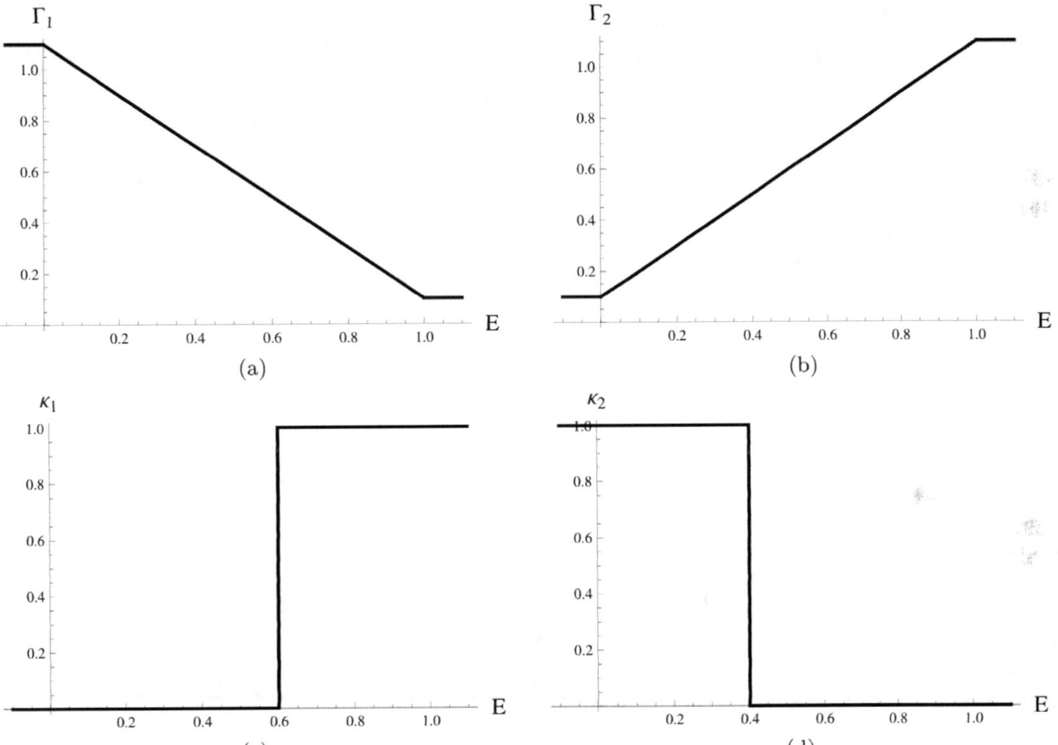

Figure 1: Functions (2) and (3).

In model (1) there is no penalty for a bacteria switching from one phenotype to another. This assumption is justified, in particular, if the time taken for the bacteria to change phenotype is shorter than the time between the occurrence of these transitions as is the case studied in [14]. The alternative case is when a cost of the transition is incorporated into a model. We make the assumption that when a bacterium begins to change phenotype it undergoes a period of shock during which it does not reproduce. To model this behaviour, we introduce two new population groups $\tilde{z}$ and $\tilde{w}$, which represent intermediate states in the transition from the phenotype $\tilde{x}$ to the phenotype $\tilde{y}$ and vice versa, respectively. That is, when a bacterium in the phenotype $\tilde{x}$ undergoes a transition to the phenotype $\tilde{y}$, it first enters the population $\tilde{z}$ and spends a period of time in this group where it does not reproduce. The bacterium then moves from the group $\tilde{z}$ to the phenotype $\tilde{y}$. The group $\tilde{w}$ works in a similar manner for transitions from the phenotype

$\tilde{y}$ to the phenotype $\tilde{x}$. This gives the new model

$$\tilde{x}' = \Gamma_1(E)\tilde{x} - \kappa_1(E)\tilde{x} + \delta_2\tilde{w}, \tag{4a}$$
$$\tilde{y}' = \Gamma_2(E)\tilde{y} - \kappa_2(E)\tilde{y} + \delta_1\tilde{z}, \tag{4b}$$
$$\tilde{z}' = \kappa_1(E)\tilde{x} - \delta_1\tilde{z}, \tag{4c}$$
$$\tilde{w}' = \kappa_2(E)\tilde{y} - \delta_2\tilde{w}, \tag{4d}$$

where $\Gamma_1(E)$ and $\Gamma_2(E)$ are given by equation (2); κ_1 and κ_2 are step functions (3) with the jumps occurring at $E = E_T + \alpha$ and $E = E_T - \alpha$, repectively; and, the parameters δ_1 and δ_2 represent the inverse of the average shock period that the bacteria undergo for changing phenotypes. In order to simulate this model numerically, we perform a change of variables and consider the population of the bacteria in each group in terms of its fraction of the total population. With the change of variables $x = \frac{\tilde{x}}{N}$, $y = \frac{\tilde{y}}{N}$, $z = \frac{\tilde{z}}{N}$ and $w = \frac{\tilde{w}}{N}$ where $N = \tilde{x} + \tilde{y} + \tilde{z} + \tilde{w}$, we obtain

$$x' = \Gamma_1(E)(1-x)x - \kappa_1 x + \delta_2 w - \Gamma_2(E)xy, \tag{5a}$$
$$y' = \Gamma_2(E)(1-y)y - \kappa_2 y + \delta_1 z - \Gamma_1(E)xy, \tag{5b}$$
$$z' = \kappa_1 x - \delta_1 z - \Gamma_1(E)zx - \Gamma_2(E)zy, \tag{5c}$$
$$w' = \kappa_2 y - \delta_2 w - \Gamma_1(E)wx - \Gamma_2(E)wy \tag{5d}$$

for the set of equations (4).

2.2. Analysis and results

The average growth rate, λ, for the system (5) is

$$\lambda = \lim_{t\to\infty}\frac{1}{t}\int_0^t (\Gamma_1(E)x + \Gamma_2(E)y)\,\mathrm{d}\tau, \tag{6}$$

where x and y are the fractional populations of the phenotypes $\tilde{x}$ and $\tilde{y}$ and $\Gamma_1(E)$ and $\Gamma_2(E)$ are the growth rates of those phenotypes. Our interest is to see the effect on the average growth rate of the system when we make changes to the thresholds at which the functions κ_1 and κ_2 change value. We only consider the case of symmetric shifts in the switching thresholds, that is if we change the threshold value by an amount α then the switching threshold of κ_1 is $E_T + \alpha$ and the threshold of κ_2 becomes $E_T - \alpha$. The value of α varies over the range $(-3, 3)$. A negative value of α means that the bacteria begin to change before the environment has stopped favouring their phenotype. Switching in this region we call *predictive* switching as the bacteria attempt to predict changes in the environment before they occur. A zero value of α means that the bacteria switch phenotype when the environment starts to favour the other phenotype marginally stronger. We call switching at this value *environmental* switching, or *memoryless* switching, as the bacteria switch each time the environment passes the point $E = E_T$. A positive value of α means that the bacteria do not switch until a sufficient bias for the other phenotype has been created in the environment. We call this switching *delayed* switching as the bacteria delay changing phenotypes until there is a strong favouring of the other phenotype. A positive α introduces memory into the system as the response of bacteria to variations of the environment depends on their state (phenotype).

The environmental state is chosen to vary as a stochastic single well potential centered at the value E_T. The environmental input to the system of equations (5) is thus given by

$$\mathrm{d}E = -(E - E_T)\mathrm{d}t + \mathrm{d}W_t, \tag{7}$$

where W_t is the Weiner process. As the mean value of E equals $E_T = 0.5$, in the absence of any stochastic fluctuations both phenotypes grow at the same rate.

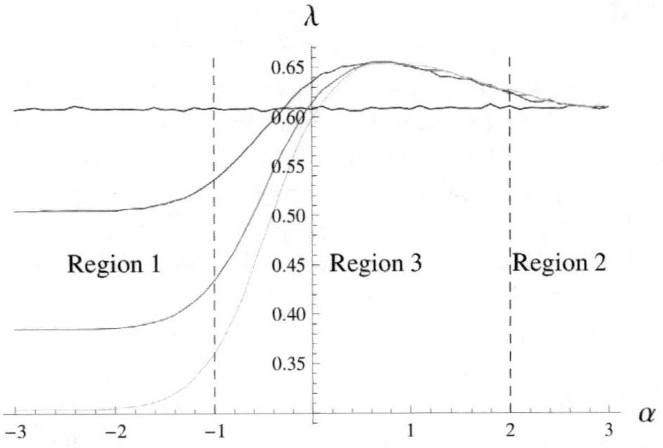

Figure 2: Plot of the average growth rate λ for various α obtained by averaging 20 simulations of the system of equations (5). The black line is for a transition rate $\kappa = 0.0$, the blue line is for $\kappa = 0.5$, the red line is for $\kappa = 1.0$, and the green line is for $\kappa = 1.5$. The other parameters used in the simulations were $\gamma_{unfit} = 0.1$, $\sigma = 1$ and $\delta_{1,2} = 1$. The initial populations of the groups were $x_0 = 0.5$, $y_0 = 0.5$, $z_0 = 0$ and $w_0 = 0$, the environment started from the state $E_0 = 0$.

Figure 2 presents results of numerical integration of system (5) with the input (7). We divide the plot into three regions. We take $\alpha \leq -1$ as region 1, $\alpha \geq 2$ as region 2 and $-1 < \alpha < 2$ as region 3. The black line gives the results for $\kappa = 0$, the case where there is no transitions between phenotypes. Here the average growth rate λ for all the values of α is nearly constant and close to the value $\gamma_{unfit} + 0.5\sigma = 0.6$, which corresponds to the growth rate at $E = E_T$ of both phenotypes. The variability of the line is a result of the stochastic nature of the input to the system and should be eliminated by the inclusion of more simulations.

In region 1, by increasing the value of κ we rapidly decrease the average growth obtained. For red, blue and green lines, the value of α is sufficient so that the environment input rarely reaches the thresholds of κ_1 and κ_2. Since α is negative and the thresholds of the functions κ_1 and κ_2 are rarely reached, the majority of bacteria are nearly always in a transition state. The higher the transition rate κ becomes the higher the percentage of the total population that gets stuck in the groups z and w which do not add to the growth rate of the system, hence lower λ.

In region 2, the value of α is also sufficient so that the environmental input rarely alters the value of the functions κ_1 and κ_2. Since in this region α is positive, both $\kappa_{1,2}$ are nearly always zero and the majority of bacteria are in the non-transition states x, y. In this region, the plots of λ for the non-zero values of κ tend to that obtained for $\kappa = 0$ as α increases.

In region 3, we see that each of the lines corresponding to a non-zero value of κ has a distinct peak average growth rate for some value of $\alpha > 0$. This means that the bacteria modelled by the system of equations (5) with the environment given by equation (7) achieve the maximum average growth rate by the phenotypes behaving under the delayed switching method. That is, they delay switching phenotypes until the environment has a strong favouring of the other phenotype. This delayed switching behaviour is beneficial to the system as a whole as it eliminates any unnecessary switching that may occur as the environment varies stochastically. The unnecessary switching is a result of the environmental input crossing the environmental threshold E_T and then quickly changing back.

3. Multi-species model

3.1. Model equations

We have shown that delaying switching between phenotypes helps bacteria to achieve a higher average growth rate in a stochastic environment in model (5). If bacteria maximise the average growth rate by having a delay of α on the environmental threshold, then, in the case of high transition rate κ, the dynamics of the state that bacteria are in can be modelled by the non-ideal relay, R_α, with thresholds $-\alpha$, α, values ± 1, and input E; for a discussion of the dynamics of the non-ideal relay see, for example, [21]. When $R_\alpha = 1$, the bacteria is in the state (phenotype) which favours the nutrient F_1; if $R_\alpha = -1$, then the bacteria favours the nutrient F_2. Adopting this model for the states of the bacteria, we consider a collection of populations of bacteria with different thresholds α sharing the same food supply. The density of bacteria with the threshold α at a moment t is denoted by $n(t, \alpha)$. We assume that a bacterium can sporadically change its threshold α. We model the resulting variability in α by a diffusion process[2]. With these assumptions, we consider the following model of evolution of bacteria and food,

$$\frac{\partial n}{\partial t} = k\Delta n + \frac{1}{2}(1 + R_\alpha(p; R_\alpha^0))nF_1 + \frac{1}{2}(1 - R_\alpha(p; R_\alpha^0))nF_2, \tag{8a}$$

$$\frac{\mathrm{d}F_1}{\mathrm{d}t} = -\frac{1}{2}F_1 \int_{\underline{\alpha}}^{\overline{\alpha}} (1 + R_\alpha(p; R_\alpha^0))n \,\mathrm{d}\alpha, \tag{8b}$$

$$\frac{\mathrm{d}F_2}{\mathrm{d}t} = -\frac{1}{2}F_2 \int_{\underline{\alpha}}^{\overline{\alpha}} (1 - R_\alpha(p; R_\alpha^0))n \,\mathrm{d}\alpha. \tag{8c}$$

Here all the non-ideal relays have the same input $p = E - E_T$ where $E = F_1/(F_1 + F_2)$, $E_T = 0.5$; the state $R_\alpha(p; R_\alpha^0)$ of the non-ideal relay R_α is a binary function of time; R_α^0 denotes the initial state of this relay; k is the rate at which the diffusion of the thresholds in the birth process occurs; $\frac{1}{2}(1 + R_\alpha)F_1 n$ is the growth rate for bacteria in the state 1; and, $\frac{1}{2}(1 - R_\alpha)F_2 n$ is the growth rate for bacteria in the state -1. Both growth rates are proportional to the population density with the coefficient of proportionality scaled to unity. The rate of consumption of food in the equation for $F_i = F_i(t)$ is proportional to the total number of bacteria in the phenotype eating this type of food, hence the integral (the coefficient of proportionality is also set to unity for simplicity); $\underline{\alpha}$ and $\overline{\alpha}$ are the lower and upper bounds on available threshold values, respectively. A certain amount of food is available at the initial moment; the food is not supplied after that moment. Bacteria do not die but stop growing when all the food has been consumed.

In model (8), we make the assumption that for a given α all the bacteria with this α value are in the same state at a particular time moment. That is, $n(t, \alpha)$ is the total population of bacteria with the threshold α at the moment t and they are all in the same state. It means that when a bacterium with a threshold α' sporadically changes its threshold to a different value α, it simultaneously copies the state from other bacteria which have the threshold α. In particular, this may require a bacterium to change the state when its threshold changes. Models where the state of a bacterium remains unchanged after a change of the threshold will be considered in a different work.

To complete the model, we assume that α values of bacteria are restricted to a selected region $\underline{\alpha} \le \alpha \le \overline{\alpha}$ and so apply the Neumann boundary conditions

$$\frac{\partial n}{\partial \alpha} = 0, \quad \alpha = \underline{\alpha}, \overline{\alpha}$$

to system (8). Replacing the variables F_1, F_2 with the new variables $F = F_1 + F_2$, the total food, and $p = E - E_T = F_1/(F_1 + F_2) - 0.5$, the deviation of the fraction of the food F_1 from

[2] Additional diffusion can occur in the birth process if we assume that a bacterium with a threshold α produces offsprings with different thresholds, for example, according to the Gaussian distribution centered at α

0.5, we obtain the equivalent system

$$\frac{\partial n}{\partial t} = k\Delta n + \frac{1}{2}Fn + pFnR_\alpha(p; R^0_\alpha), \tag{9a}$$

$$\frac{\mathrm{d}F}{\mathrm{d}t} = -\frac{1}{2}F\int_{\underline{\alpha}}^{\overline{\alpha}} n\,\mathrm{d}\alpha - pF\int_{\underline{\alpha}}^{\overline{\alpha}} nR_\alpha(p; R^0_\alpha)\,\mathrm{d}\alpha, \tag{9b}$$

$$\frac{\mathrm{d}p}{\mathrm{d}t} = (p^2 - \frac{1}{4})\int_{\underline{\alpha}}^{\overline{\alpha}} nR_\alpha(p; R^0_\alpha)\,\mathrm{d}\alpha. \tag{9c}$$

The last integral in the second equation and the integral term in the last equation are the Preisach operator with the input $p = p(t)$ and the time dependent density function $n(t, \alpha)$ concentrated on the bisector of the Preisach half-plane [21]. Well-posedness of this system has been established in [22].

3.2. Results for the multi-species model

The distribution of phenotypes for different threshold values after long time simulation of system (9) is shown in Figures 3 and 4. In each simulation, the distribution, which at any point in time is a binary function of α, stabilises to some stationary limit profile. Initially, all the non-ideal relays were set to be in the state $R^0_\alpha = 1$, $\underline{\alpha} \le \alpha \le \overline{\alpha}$. The initial population density $n(0, \alpha)$ was set to a non-zero value at the point $\alpha = \overline{\alpha}$ and to zero in all the other nodes of the α-mesh; that is, all the population initially consisted of bacteria with the largest threshold. The food supply was equally distributed between the food types F_1 and F_2, that is the initial value of p was $p(0) = 0$.

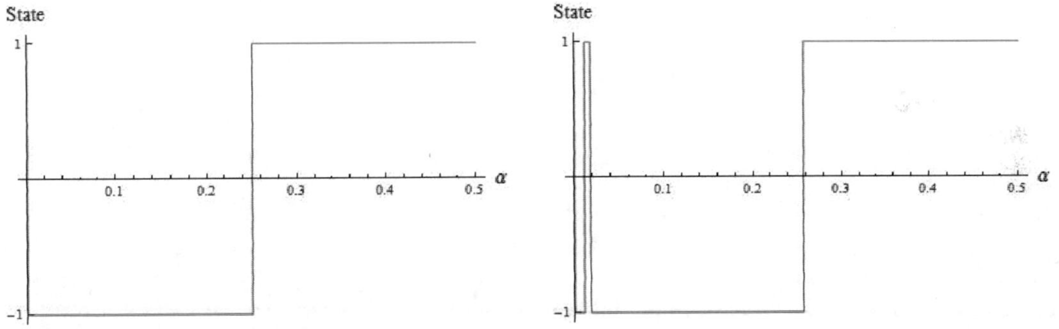

(a) The initial food supply $F(0) = 10$ and the diffusion rate $k = 0.1$.

(b) The initial food supply $F(0) = 100$ and the diffusion rate $k = 0.1$.

Figure 3: Stationary limit distribution of the states of the relays after long time simulation of the system (9) in case of faster diffusion. The range of available thresholds is $[\underline{\alpha}, \overline{\alpha}] = [0, 0.5]$.

By comparing Figures 3a and 4a, we see that with the decrease of the diffusion rate, k, the limit distribution takes on a different form. In the case of faster diffusion between populations with different α, the phenotypes split equally into two regions, those in the region $0 \le \alpha \le 0.25$ are in the state -1 and those in the region $0.25 < \alpha \le 0.5$ are in the state 1. In the case of slower diffusion, we observe a more oscillatory pattern in the limit distribution. In both cases, the density n converges to the same constant value.

Figures 3b and 4b show the changes in the distribution caused by increasing the initial food supply. Increasing the initial food supply increases the number of oscillations in the distribution of the relays. Decreasing the diffusion rate has a greater effect on causing the distribution to take on an oscillatory pattern.

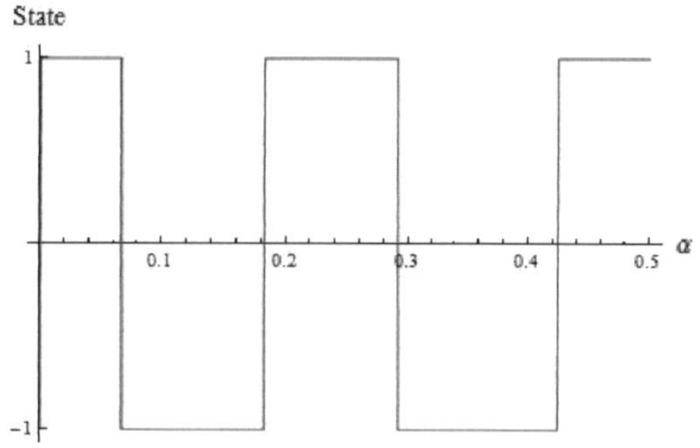

(a) The initial food supply $F(0) = 10$ and the diffusion rate $k = 0.001$.

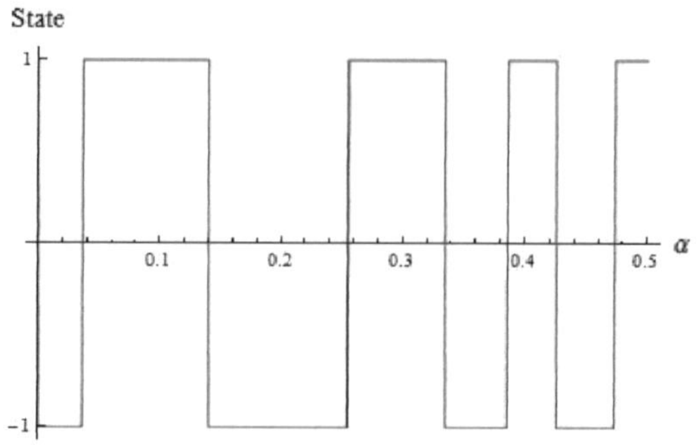

(b) The initial food supply $F(0) = 100$ and the diffusion rate $k = 0.001$.

Figure 4: Stationary limit distribution of the states of the relays after long time simulation of the system (9) in case of slower diffusion. The range of available thresholds is $[\underline{\alpha}, \overline{\alpha}] = [0, 0.5]$.

4. Conclusions

We developed a model for two phenotype bacteria with a cost for transitioning between phenotypes and saw that, in an environment that varies stochastically, the average growth rate of the system could be maximised by the bacteria delaying the phenotype switching until there was a stronger favouring of the other phenotype. Dynamics of the optimal system can be interpreted as a kind of a probabilistic memory switch, or relay, driven by a varying input (environment). The standard deterministic non-ideal relay R_α switches from state -1 to state 1 when the input reaches the threshold value α, and switches back when the input achieves the value $-\alpha$; transitions in both directions are instantaneous. We define the switching rules for the probabilistic relay as follows. The relay does not switch from state -1 to state 1 as long as the input satisfies $p < \alpha$. When $p > \alpha$, we assume a positive transition probability rate κ for transitioning from -1 to 1 (that is, the exponential distribution of the resident time in state -1 before switching, as in the Poisson process). Similarly, there is no switching from state 1 to -1 as long as $p > -\alpha$; and, the transition probability rate $\kappa > 0$ when $p < -\alpha$.

In model (5), we also have an intermediate state, so that when the relay departs from state -1 (or, 1) , it spends a period of time in the intermediate state before arriving in state 1 (state -1, respectively). A similar behaviour of the probabilistic relay can be achieved by adding two intermediate states A and B; assuming the transition probability rate $\kappa > 0$ from state -1 to A when $p > \alpha$ and zero probability rate from -1 to A for $p < \alpha$; assuming a positive probability rate δ_1 from A to 1 (independent of the value of p); and, similarly, assuming the transition probability rate κ from state 1 to B when $p < -\alpha$, zero probability rate from 1 to B for $p > -\alpha$, and a positive probability rate δ_2 from B to -1 independent of p. With this modification, the evolution of the probabilities to find the probabilistic non-ideal relay in states $-1, 1, A, B$ is described by deterministic equations similar to (5).

The above dynamics, which is achieved in model (5) for positive α, can maximise the growth rate of the population of bacteria in a stochastic environment if a typical time that bacteria spend in the intermediate states is comparable to the characteristic time of the input variation. A positive delay α prevents transitions from less to more favoured phenotype when favouring is not strong. A slight decrease in the growth rate due to small fluctuations of the environment from the point E_T can be less dramatic than a drop in the growth rate due to passing through the intermediate states where bacteria do not reproduce. This trade off between too much responsiveness (small α), with the associated cost of often transitions, and too much inertia (large α), which leaves too many bacteria in unfavoured states, shifts α_{opt} to the positive range

Journal of Physics: Conference Series **585** (2015) 012012 doi:10.1088/1742-6596/585/1/012012

provided that the time bacteria have to spend in the intermediate states is significant.

If transitions between the states are fast compared to the time scale of input variations (large κ and $\delta_{1,2}$), then dynamics of system (5) can be approximated by the deterministic non-ideal relay R_α. For example, this is the case if we consider a population of bacteria which are best fit to some stochastic environmental input with $\alpha \approx \alpha_{opt}$ in model (5), and then change the input to a much slower, less variable, or more predictable environment (such as, for instance, in a lab). In the new environment E_{slow}, nearly all the population, at almost any given instant on the slow time scale, will have the same phenotype; transitions between the phenotypes will be almost instantaneous; and, switching between the two phenotypes will be described by the deterministic non-ideal relay R_α with the input E_{slow}; that is, the system will demonstrate persistent bistable (hysteretic) memory.

Assuming this simpler switching regime, we considered a reaction-diffusion model of multiple populations of bacteria competing for food. The switching rules for each population were modelled by the deterministic relay R_α; the populations were parametrised by the thresholds α varying over an interval of values. In our model, a finite amount of food is available, bacteria stop growing but do not die when all the food has been consumed, and the population density $n(t, \alpha)$ tends to a positive constant as t increases. We have shown numerically that a binary function describing the distribution of two phenotypes among bacteria with different α also converges to a stationary pattern. However, this sign changing pattern is different for different initial data. The limit stationary pattern becomes more oscillatory with the increase of the initial food supply. The limit pattern is sensitive to the diffusion rate between species with different thresholds. Lowering the diffusion rate leads to more oscillatory patterns.

It would be interesting to consider variations of the multi-species model (9) and their effect on the attractor and the pattern formation. Possible modifications might account for the death process and constant or variable food supply; testing the model with no diffusion process; the inclusion of relays $R_{\alpha,\beta}$ with asymmetric thresholds α, β, $\beta \neq -\alpha$; variations of the boundary conditions; replacing each deterministic relay with the differential model (5); or, modelling populations by a collection of probabilistic memory switches using stochastic differential equations. Systems of reaction-diffusion equations coupled to ordinary differential equations with bistable dynamics have been used in [23, 24] to model the formation of spatial patterns observed experimentally in a culture of bacteria grown in a Petri dish. An important assumption we made in multi-species model (9) was that bacteria, when sporadically changing their threshold α' to a new value α, simultaneously copy the state from their peers who have the same threshold α. It would be natural to explore a model where the state remains unchanged when the threshold changes. Such a model should have simultaneous nonzero populations of bacteria with the same threshold in two phenotypes. This is a subject of future work.

The nonlocal terms introduced by the integral $\int_{\underline{\alpha}}^{\overline{\alpha}} nR_\alpha(p; R_\alpha^0)\, d\alpha$ in equations (9) is the Preisach operator with the input p and the time dependent density function $n(t, \alpha)$. The evolution of the Preisach density function is coupled to the evolution of other variables in the system. This contrasts to all the variety of known applications of the Preisach operator model, where one needs to know the Preisach density function *a priori* in order to parametrise the model [25] and analyze dynamics (see, for example, [26–37]). Identification of the Preisach density function is a daunting task requiring multiple experiments in applied physics and engineering [38]; measurements of the density function in applications to economics and finance are even less accessible. In model (9) and its modifications, the profile of the density function at any moment is defined by its initial profile $n(0, \alpha)$. However, if one is interested in long term behaviour, then no (or, little) information about the initial profile is needed provided that the attractor has a simple structure. That is, in this case, one does not need to solve the problem of identification of the Preisach operator density function. On the other hand, we have shown that the diffusive evolution of the Preisach density function leads to a new mechanism of pattern

Journal of Physics: Conference Series **585** (2015) 012012 doi:10.1088/1742-6596/585/1/012012

formation, which is not observed in systems with the steady Preisach density functions.

Acknowledgments

DR acknowledges the support of the Russian Foundation for basic Research through grant 10-01-93112.

References

[1] D M Wolf, L Fontaine-Bodin, I Bischofs, G Price, J Keasling and A P Arkin 2008 *PLoS One* **3** 2 e1700
[2] D E Koshland Jr 1977 *Science* **196** 10551063
[3] B Hallet 2001 *Curr Opin Microbiol* **4** 570581
[4] N J Holden and D L Gally 2004 *J Med Microbiol* **53** 585593
[5] D Dubnau and R Losick 2006 *Mol Microbiol* **61** 564572
[6] O A Igoshin, C W Price and M A Savageau 2006 *Mol Microbiol* **61** 165184
[7] D M Wolf and A P Arkin 2003 *Curr Opin Microbiol* **6** 125134
[8] A Arkin, J Ross and H H McAdams 1998 *Genetics* **149** 16331648
[9] T S Ham, S K Lee, J D Keasling and A P Arkin *2008 PLoS One* **3** 7 e2815
[10] A Novick and M Weiner 1957 *Proc Natl Acad Sci USA* **43** 553566
[11] J E Ferrell Jr 2002 *Curr Opin Cell Biol* **14** 140148
[12] S R Biggar and G R Crabtree 2001 *EMBO J* **20** 31673176
[13] M Thattai and B Shraiman 2003 *Biophys J* **85** 744754
[14] M Thattai and A van Oudenaarden 2004 *Genetics* **167** 523-530
[15] E Kussell and S Lieber 2005 *Science* **309** 2075-2078
[16] S M Hoffer, H V Westerhoff, K J Hellingwerf, P W Postma and J Tommassen 2001 *J Bacteriol* **183** 49144917
[17] R Aumann and M Maschler 1995 *Repeated Games with Incomplete Information* (Cambridge: MIT Press)
[18] M Nowak and K Sigmund 1993 *Nature* **364** 5658
[19] R Axelrod and W D Hamilton 1981 *Science* **211** 1390-1396
[20] D M Wolf, V J Vazirani and A P Arkin 2005 *J Theor Biol* **234** 2 227-253
[21] M A Krasnosel'skii and A V Pokrovskii 1989 *Systems with Hysteresis* (Springer)
[22] P Gurevich and D Rachinskii *Proc Steklov Inst Math* **283** 1 87-109
[23] C Chiu, F C Hoppensteadt and W Jäger 1994 *J Math Biol* **32** 841-855
[24] C Chiu and N Walkington 1998 *Quarterly of Applied Mathematics* **LVI** 1 89-106
[25] *The Science of Hysteresis* 2005 I Mayergoyz and G Bertotti eds (Elsevier, Academic Press)
[26] B Appelbe, D Flynn, H McNamara, P O'Kane, A Pimenov, A Pokrovskii, D Rachinskii and A Zhezherun 2009 *IEEE Control Syst Mag* **29** 1 44-69
[27] B Appelbe, D Rachinskii and A Zhezherun *Physica B* **403** 2 301-304
[28] M Brokate, A Pokrovskii, D Rachinskii and O Rasskazov 2005 *The Science of Hysteresis* **I** I Mayergoyz and G Bertotti eds 125-291 (Elsevier, Academic Press)
[29] M Brokate, A Pokrovskii and D Rachinskii 2006 *J Math Anal Appl* **319** 1 94-109
[30] R Cross, H McNamara, A Pokrovskii and D Rachinskii 2008 *Physica B* **403** 2 231-236
[31] P Diamond, N Kuznetsov and D Rachinskii 2001 *J Differ Equations* **175** 1 1-26
[32] P Diamond, D Rachinskii and M Yumagulov 2000 *Nonlinear Anal - Theor* **42** 6 1017-1031
[33] A Krasnosel'skii and D Rachinskii 2002 *NoDEA Nonlinear Differential Equations Appl* **9** 1 93-115
[34] A M Krasnosel'skii and D I Rachinskii 2001 *Doklady Mathematics* **63** 3 339-344
[35] P Krejci, J P O'Kane, A Pokrovskii and D Rachinskii 2011 *J Phys: Conf Ser* **268** 1 012016
[36] A Pimenov and D Rachinskii 2009 *Discrete Cont Dyn Sys B* **11** 997-1018
[37] D Rachinskii 1999 *NoDEA Nonlinear Differential Equations Appl* **6** 3 267-288
[38] I D Mayergoyz 1991 *Mathematical Models of Hysteresis* (Springer)

A preliminary threshold model of parasitism in the Cockle *Cerastoderma edule* using delayed exchange of stability

E A O'Grady[1], S C Culloty[2], T C Kelly[2], M J A O'Callaghan[1], D Rachinskii[1,3]

[1] Department of Applied Mathematics, University College Cork, Ireland
[2] School of Biology, Earth and Environmental Sciences, University College Cork, Ireland
[3] Department of Mathematical Sciences, The University of Texas at Dallas, Richardson, USA

E-mail: dmitry.rachinskiy@utdallas.edu

Abstract. Thresholds occur, and play an important role, in the dynamics of many biological communities. In this paper, we model a persistence type threshold which has been shown experimentally to exist in hyperparasitised flukes in the cockle, a shellfish. Our model consists of a periodically driven slow-fast host-parasite system of equations for a slow flukes population (host) and a fast *Unikaryon* hyperparasite population (parasite). The model exhibits two branches of the critical curve crossing in a transcritical bifurcation scenario. We discuss two thresholds due to immediate and delayed exchange of stability effects; and we derive algebraic relationships for parameters of the periodic solution in the limit of the infinite ratio of the time scales. Flukes, which are the host species in our model, parasitise cockles and in turn are hyperparasitised by the microsporidian Unikaryon legeri; the life cycle of flukes includes several life stages and a number of different hosts. That is, the flukes-hyperparasite system in a cockle is, naturally, part of a larger estuarine ecosystem of interacting species involving parasites, shellfish and birds which prey on shellfish. A population dynamics model which accounts for one system of such multi-species interactions and includes the fluke-hyperparasite model in a cockle as a subsystem is presented. We provide evidence that the threshold effect we observed in the flukes-hyperparasite subsystem remains apparent in the multi-species system. Assuming that flukes damage cockles, and taking into account that the hyperparasite is detrimental to flukes, it is natural to suggest that the hyperparasitism may support the abundance of cockles and, thereby, the persistence of the estuarine ecosystem, including shellfish and birds. We confirm the possibility of the existence of this scenario in our model, at least partially, by removing the hyperparasite and demonstrating that this may result in a substantial drop in cockle numbers. The result indicates a possible significant role for the microparasite in this estuarine ecosystem.

1. Introduction

Parasites are now widely recognised to play an important role in the structuring and functioning of ecological communities. In the estuarine systems of the south western USA, for example, Kuris *et al.* (2008) showed that parasite biomass "exceeded that of top predators" while the biomass of trematodes "equalled" that of the most abundant birds, fish and crustacea. Macroparasites, such as digenetic trematodes (also commonly known as flukes), have complex life cycles with several intermediate hosts and complex transmission dynamics to their final, usually vertebrate, hosts. Although it has been known for some time that the flukes themselves may be parasitized by microparasites, the general importance of this relationship in the population dynamics of the interacting species has only recently been recognised (Raffel *et al.* 2008). The dynamics of many host-parasite relationships are subject to thresholds of several different types (Getz and Pickering, 1983; Heesterbeek and Roberts, 1995; Deredec and Courchamp, 2003; Kalachev *et al.*, 2011). In this section, we model the impact of a "persistence" type threshold in a hyperparasitised fluke which infects a common estuarine bivalve mollusc – the cockle *Cerastoderma edule.* A persistence-type threshold is relevant to parasites and is characterised by a necessary number or density of hosts for parasite maintenance (*sensu* Deredec and Courchamp, 2003). We begin by describing in more detail the main species involved in the study.

1.1. The Cockle

The cockle *Cerastoderma edule* populates estuarine regions from the Barents Sea to Morocco, including the coastlines of Britain and Ireland, the area in which this study was undertaken. Not only is this species of commercial importance, but it also plays a significant role in estuarine ecosystems as an important food source for birds, fish and crustaceans as well as in the dynamics and stability of the sediment. Cockle densities at different sites can be highly variable with 1000 per square metre not unusual but as many as 54, 400 being recorded (Ducrotoy *et al*, 1991). A range of variables control cockle population dynamics, including biotic influences such as bioturbation (that is burrowing activity and aeration of sediment), predation, parasitism and food availability together with abiotic factors such as temperature, immersion time, water velocity and sediment dynamics (Gam *et al.*, 2010). In particular, recruitment is significantly affected by predation by species such as the shrimp *Crangon crangon* (see for example, Beukema and Dekker, 2005).

1.2. The Macroparasite - the fluke

In intertidal ecosystems, flukes are the dominant parasite group (Mouritsen and Poulin, 2002). These macroparasites exert two influences; firstly as part of the living diversity, and secondly as diversity indicators, because their presence is linked to the richness of the free living fauna (their hosts) as demonstrated by Hechinger *et al.* (2007), thereby acting as a proxy for ecosystem health, (Hudson *et al.* 2006). These digenetic trematodes generally have three hosts in their life-cycle: a definitive host where the parasite reaches sexual maturity, a molluscan first intermediate host in which asexual reproduction occurs, and a second intermediate host which provides the vehicle for transfer back to the definitive host (Esch *et al.*, 2002).

Cockles are first or second intermediate hosts of, at least, 16 fluke species with complex life cycles involving 2-3 hosts (de Montaudouin *et al.*, 2009; Russell-Pinto *et al.*, 2006). This diversity in parasite infracommunities seems to be possible due to the observed high degree of spatial segregation of parasite species within their cockle host (Rohde, 1994). One such fluke is *Meiogymnophallus minutus*, which is one of four sibling species, all of which infect cockles of the genus *Cerastoderma.* The fluke can be found in the mantle epithelium of the cockle (Bowers *et al.*, 1996; de Montaudouin *et al.*, 2009). The fluke has a body length of 240-350 μm and a width of about 150 μm (Russell-Pinto, 1990). This trematode is a member of the Family Gymnophallidae, with the bivalve mollusc *Scrobicularia plana* being the first intermediate host, *C.edule* the second intermediate host, and the oystercatcher *Haematopus ostralegus* - a widely distributed shorebird - the final, or definitive, host. This cycle is described schematically in Figure 1.

It has been demonstrated that larval stages of *M. minutus* released from the first intermediate host and dispersed via water currents, passively locate their second intermediate host within a distance of a few hundred metres, where they encyst becoming metacercariae. These do not reproduce and therefore population growth within the cockle is solely due to the incremental effects of immigration. The impact of *M. minutus,* when present in large numbers, is to inhibit the tight closing of the shell valves. Consequently, these cockles are often found, shell gaping upwards, close to the surface of the sand (Bowers *et al.*, 1996). Cockles infected with flukes suffer from a range of different effects such as impaired burrowing, reduced growth, increased mortality and reduced tolerance to opportunistic microparasitic infections (Lauckner *et al.*, 1983; Wegeberg and Jensen, 1999; 2003).

In a study conducted along the Northeast Atlantic, this fluke was found to be the most abundant and widespread trematode in cockles (de Montaudouin *et al.*, 2009) from the Wadden Sea in the north to Morocco in the south (Gam *et al.*, 2008). This distributional range is linked to large scale migrations of shorebirds (Thieltges and Reise, 2006). In previous studies the prevalence of *M. minutus* had been found to range from up to 48% in the Wadden Sea, (Thieltges and Reise, 2006), to as much as 100% in Arcachon Bay, France, (de Montaudouin *et al.*, 2000), in Merja Zerga in Morocco (Gam *et al.*, 2008) and at a number of sites along the south coast of Ireland (Fermer *et al.*, 2011).

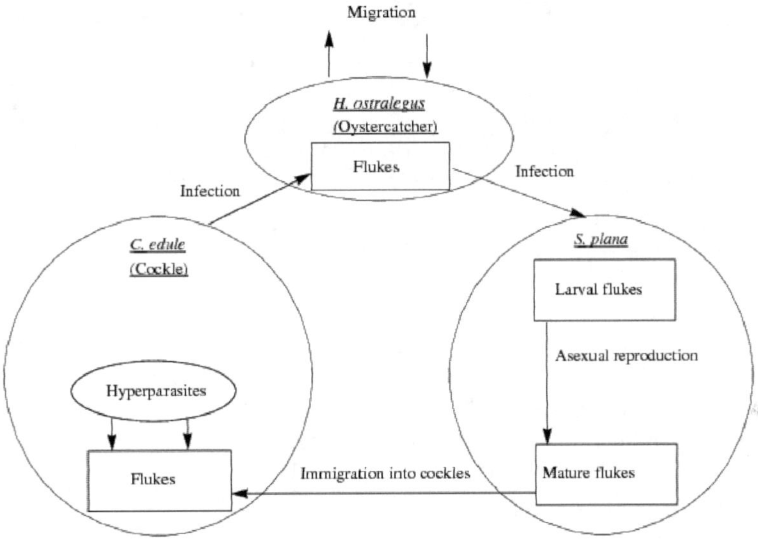

Figure 1. Flow chart showing the interactions of species involved and the life-cycle of the fluke *M. minutus*.

1.3. The Hyperparasite

Microsporidia are obligate intracellular eukaryotic microparasites, ranging in size from about 3 to 6 μm, most closely related to the Fungi (Troemel *et al.*, 2008). It is clear that they are up to two orders of magnitude smaller than the fluke. They infect both vertebrates and invertebrates and include a number of species that are found in humans (Didier *et al.*, 2004). Johnson *et al.* (2010) describe eight microsporidians, including *Unikaryon legeri* – the subject of this study, that are hyperparasites of various digenean trematodes which they classify as predators of the flukes. Although not germane to this study, the issue of whether *U. legeri* is a parasite (i.e. hyperparasite) or a predator needs to be addressed by future research because of the implications for the population dynamics of the interacting species.

1.4. Interaction between M. minutus *and* U. legeri

Spatial variation in the presence of the hyperparasite, *U. legeri,* was found by Fermer *et al.* (2010) when cockles ($N = 50$ at each site) were screened at 14 separate locations on the south coast of Ireland in July 2007. Hyperparasitism occurred at 8 out of the 14 sites and was prevalent only in cockles heavily infected with the fluke. Of particular significance was the fact that *U. legeri* was not present where the burden of *M. minutus* was less than approximately 185 metaceriae per cockle (Fermer, 2009). This low burden of metacercarial flukes within the cockle appears to be a threshold enabling infection with the hyperparasite. It is clearly evident therefore, that the threshold is that of a persistence type as defined above. It will ultimately determine the duration of infection, as seasonal effects will impact strongly on the average numbers of flukes in the cockles due to their temporal life-cycle.

It has been shown that hyperparasitism results in the death of a high proportion of the infected flukes. Due to hyperparasite-induced mortality, the entire metacercarial population has to rebuild every year. Hyperparasitism is thus a very important factor controlling fluke infrapopulation size and the overall dynamics of *M. minutus.*

1.5. Structure of the paper

In the next section we present and discuss a model of the hyperparasitised flukes in a cockle. This model is then incorporated as a subsystem in a larger ecosystem model, which is formulated and tested in Section 3. The multiscale analysis technique we use is close the method described by Nizette *et al.* (2006). The last section contains conclusions.

2. Fluke-Hyperparasite model

We present a mathematical model which describes the subsystem occurring in the individual cockle. In particular, we model the threshold population of flukes inside a cockle that is needed to provide a sufficient environment for the parasite population to grow and persist. As explained above, when fluke populations are less than 185 per cockle (Fermer, 2009), there are no hyperparasites present. On exceeding this threshold number of flukes, there is a significant change in the dynamics of both populations whereby the hyperparasite population grows, attacks and can kill part of the fluke population. The parasite has a regulatory effect on the fluke population which appears to be unique (see Raffel *et al.*, 2008).

2.1. Threshold model development

Flukes enter the cockle through immigration as outlined above. We assume a natural death rate of flukes and an interaction between the flukes and the hyperparasites that we consider to be of a parasitic type. On the basis of these assumptions, we write the following equation for the rate of change of the fluke population:

> Rate of change of fluke population = (Immigration term) − (Natural death rate) − (Death due to hyperparasites),

which leads to the equation:

$$\frac{dF}{dt} = a(t) - \delta F - \lambda H,$$

where F is the fluke population, H is the hyperparasite population, the derivative dF/dt represents the rate of change of fluke population and $a(t)$ represents a time-dependent immigration rate. The term $-\lambda H$ represents the death rate of flukes due to the hyperparasites. As a parasitic type term, we assume that the damage caused by the hyperparasite to a single fluke is proportional to the mean number that the fluke carries, that is, H/F. Thus the decrease in the overall fluke population will be proportional to H. Here λ is a positive constant representing the number of flukes lost per hyperparasite per unit time.

Journal of Physics: Conference Series **585** (2015) 012013 doi:10.1088/1742-6596/585/1/012013

For the hyperparasite population, we assume a natural death rate in the absence of flukes, a positive growth term due to the exploitation of their hosts, and a compensating term to reflect the damage that host mortalities have on the population:

> Rate of Change of parasite population = − (Death rate) + (Growth term due to parasitism on flukes) − (Compensating term).

The second, growth, term here was developed as follows. The rate of production of hyperparasite transmission stages is proportional to H. The magnitude of the proportion of these hyperparasites that are successful in infecting flukes is characterised by a term reflecting the relative density of hosts, $F/(F_0 + F)$, where F_0 is a measure of the transmission efficiency, and F is the fluke population, as described in general terms in Anderson and May (1978). A large F_0 indicates that this efficiency is low and is the one that we use. Under this assumption, that only a small proportion of the hyperparasite transmission stages actively infect flukes, the growth term in the hyperparasite population is proportional to HF.

The compensating term reflects the fact that as it kills its host, a certain proportion of hyperparasites are also lost. An important factor needing consideration here is the distribution of hyperparasites in the flukes. Consequently, we note that as hyperparasites damage the fluke population at a rate proportional to H, they themselves will suffer loss proportional to H multiplied by a proportion of the average distribution H/F. We introduce a constant of proportionality f, to indicate this distribution proportion of hyperparasites on flukes. We will describe this coefficient in greater detail below. The equation for the rate of change of hyperparasites can then be written as:

$$\frac{dH}{dt} = -cH + dHF - \frac{f\lambda H^2}{F}.$$

It is common for systems to have dynamics that interact on differing timescales (see for example, Etchechoury and Muravchik; 2003) as described already, and so we now introduce one important assumption to this last equation. We will require that the interval over which the hyperparasite population replicates and undergoes transmission occurs on a timescale much shorter than the corresponding interval of the fluke population. This is reasonable given the fact that the hyperparasites are several orders of magnitude smaller than the flukes as mentioned above. To account for this mathematically, we introduce a new parameter ε in Eq. (2.1) and assume $0 < \varepsilon \ll 1$. Therefore we obtain the equations:

$$\frac{dF}{dt} = a(t) - \delta F - \lambda H, \tag{1}$$

$$\varepsilon\frac{dH}{dt} = -cH + dHF - \frac{f\lambda H^2}{F}. \tag{2}$$

The last term in Eq. (2) can be used to model the occurrence of over-dispersion of hyperparasites on flukes. Over-dispersion means that the majority of flukes carry few numbers of hyperparasites, while a minority are greatly infected. Thus, the death of a fluke results in a greater than average proportion of hyperparasites being lost. In this last term, we are concerned with those hyperparasites living on, and killing flukes, and we reason that the interaction occurs on timescales of that of the fluke. Thus, when we wish to examine cases of over-dispersion, it is meaningful to examine the value of the combination, $f\lambda/\varepsilon$. In particular, values of this ratio greater than one indicate over-dispersion.

Comparing our model with that described by Anderson and May (1978), it can be seen that the equation for our host, the fluke, equation is similar except that we have an immigration term that has a seasonal dependence instead of a natural birth rate, making the dynamics different. In the hyperparasite equation, we have an identical death rate and our birth rate

Table 1. Parameter values and units used in the model.

Parameter	Description	Value
a	Fluke immigration	20
δ	Fluke natural death rate	1/25
λ	Fluke death due to hyperparasitism	1/800
ε	Timescale separation parameter	1/5000
c	Hyperparasite natural death rate	185 d
d	Growth rate of Hyperparasite	1/3500

term is comparable in the regime where only a small fraction of the hyperparasite population is successful in infecting the flukes as discussed above. The compensating term is equivalent to Anderson and May's term for host-death induced parasite mortality, although written here in a different manner. We also have the notable difference of the differing time scales as introduced through the small parameter ε. This will be shown to have a major impact on the dynamics of the system.

Table 1 shows typical parameter values used in the model. Although direct experimental measurement of parameters is difficult, we can reason that there is a basis for the order of magnitude of the values used here. We use a reasonable monthly immigration rate of 20 here, particularly to examine the interesting behaviour of the threshold phenomenon. Natural death rates of both flukes and hyperparasites, δ and c respectively, are usually much less than 1, and we require that $c > \delta$ due to the fact that host deaths result in large numbers of parasite deaths. The parameter d is taken to be small here, to prevent too large a population of hyperparasites. The fluke death rate due to hyperparasitism, λ, is chosen here so that the hyperparasites have an impact on fluke population and regulate their growth. In the case of over-dispersion of hyperparasites, we want $f\lambda/\varepsilon$ greater than one and for our simulations we take a value of 5. We will see in the next section that the ratio c/d marks a point where the hyperparasite population goes to zero, or to a non-zero population, giving immediately the value used for c, if we take this to be the threshold population and assume it to be 185 (shown below).

2.2. Preliminary discussion

Two equilibria are possible in the system of Eqs. (1) and (2) with constant immigration term $a(t) \equiv a$, namely zero- and non-zero hyperparasite population equilibria, dependent on the parameters and the threshold effect we wish to model. Taking, a, the constant rate of immigration of flukes into the cockle, as a bifurcation parameter, a transcritical bifurcation can be shown to occur at $a = c\delta/d$. This transcritical bifurcation is evident as shown by Figure 2. A transcritical bifurcation indicates that a stable positive solution branches from the zero hyperparasite solution at the bifurcation point. In the current case, the zero hyperparasite equilibrium becomes unstable after the transcritical bifurcation while the positive, non-zero, hyperparasite population becomes stable. Thus, solutions after the transcritical bifurcation are attracted to the non-zero trajectory, while solutions before the transcritical bifurcation tend to the stable zero hyperparasite equilibrium. Therefore, the zero-hyperparasite equilibrium is stable if the condition $a/\delta < c/d$ is satisfied. The positive equilibrium exists and is stable if the opposite inequality holds. This figure is the same for all values of the time scale separation parameter ε. We see by Table 1, that both sides of the inequality have units of flukes. Thus we recognise that the number of flukes determines the stabilities of the zero and non-zero hyperparasite populations, and thereby, the dynamics of the system. In a real-life system, the immigration

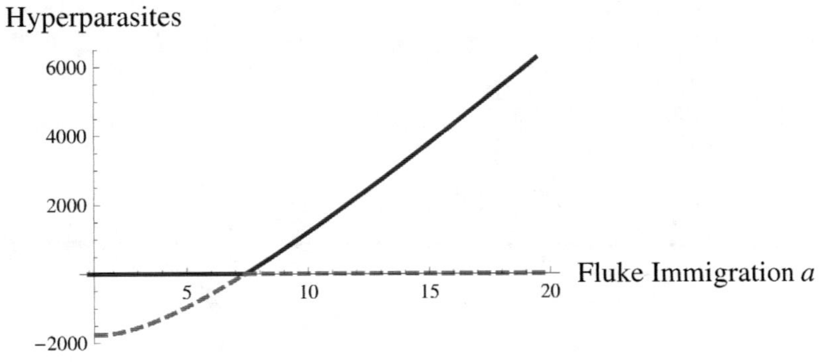

Figure 2. Bifurcation diagram showing fluke immigration rate parameter a on the horizontal axis with hyperparasite population on the vertical axis. Transcritical bifurcation point evident at $a = 7.4$, where for this parameter set $(a/\delta) = (c/d)$. Here, the previously stable zero hyperparasite solution exchanges stability with the previously unstable non-zero solution. It is evident that for this parameter set, hyperparasites are absent for a less than 7.4 and present above this value. Blue solid lines indicate stable solutions, red-dashed lines indicate unstable solutions. Here, the parameter values, apart from the bifurcation parameter a, are given in Table 1.

term varies temporally due to migratory patterns of birds who consume the cockles infected with parasites, which ultimately control the cycle of fluke reproduction, and also due to temperature variation and perhaps other abiotic influences. If $a(t)$ changes abruptly between two values, we obtain a switching system forced by a square wave function. Such a type of function to model the temporal influx of a species has been used previously in, for example, Holt *et al.* (2003). For example, at certain times of the year, the system can be attracted towards the zero-hyperparasite equilibrium due to no fluke immigration and for the rest of the year towards the positive equilibrium, shown in Figure 3.

2.3. Singular perturbation and stability exchange

First we examine the condition for the bifurcation point evident in system (1), (2).

Taking this system with constant immigration a, we wish to examine the stability of the equilibrium $(F^*, H^*) = (a/\delta, 0)$. The jacobian matrix of this system is

$$\begin{pmatrix} -\delta & -\lambda \\ 0 & (-c + dF)/\varepsilon \end{pmatrix}.$$

Hence, the eigenvalues are

$$\mu_1 = -\delta, \qquad \mu_2 = (-c + dF)/\varepsilon.$$

For stability of the equilibrium $(F^*, H^*) = (a/\delta, 0)$, we require $\mu_1, \mu_2 < 0$ (both eigenvalues are real in this case). Since μ_1 is negative by definition ($\delta > 0$), the equilibrium is stable for $c > dF^* = da/\delta$, with the transcritical bifurcation occurring at

$$\frac{a}{\delta} = \frac{c}{d}.$$

For $\varepsilon \ll 1$, Eqs. (1) and (2) become a singularly perturbed system with the hyperparasite being the fast variable (see above and, for example, Veliov, 1997). Systems of this kind, where the small parameter ε tends to zero, lend themselves to analysis of a reduced initial value problem,

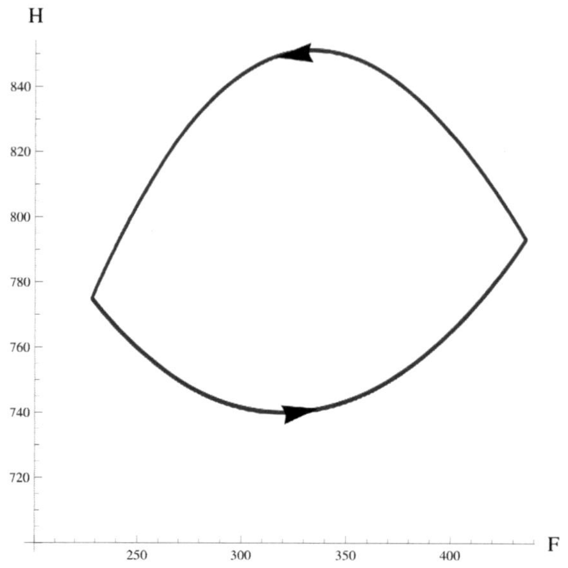

Figure 3. We include for illustrative purposes the stable periodic regime of system (1), (2) with the fluke immigration rate a switching periodically between the values $a = 0$ and $a = 70$. Here $\varepsilon = 1$, i.e. both populations evolve on the same timescale, $c = 185d$, $f = 0.9$, $\delta = 1/25$, $d = 0.00031$ and $\lambda = 1/50$. It can be seen that when immigration switches on, there is attraction towards the non-zero equilibrium, at least after the threshold number of flukes is reached. When immigration ceases, the system turns back and evolves towards the zero equilibrium, now stable at this time.

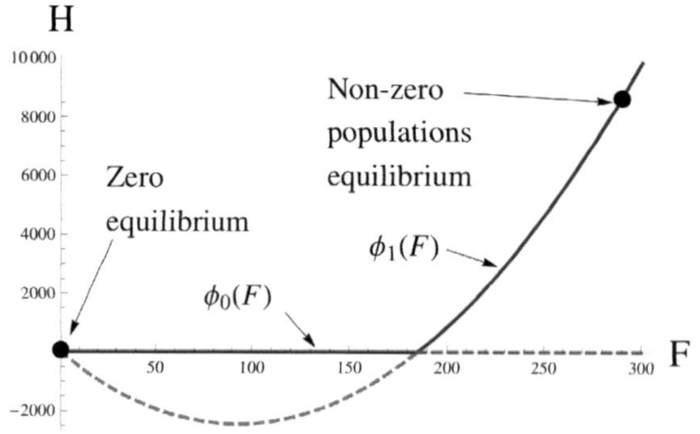

Figure 4. Intersecting critical curves $H = \phi_0(F)$, $H = \phi_1(F)$ of system (1) and (2) with parameter values shown in Table 1. Solutions of the dynamical system evolve on these curves, depending on stability of the equilibria. Two equilibria for the populations are indicated, and these depend on the value of a. When $a = 20$, immigration is switched on and solutions tend to the non-zero equilibrium. In this case, the zero equilibrium becomes unstable. When immigration is switched off, $a = 0$, the system tends towards the stable zero equilibrium.

which is one-dimensional in this case. Indeed, the trajectories of the singularly perturbed system spend most of the time near the critical manifold, actually a curve in this case, which is defined by setting the right hand side of the fast hyperparasite equation to zero, yielding the two roots

$$H = \phi_0(F) \equiv 0, \qquad H = \phi_1(F) = F(dF - c)/(f\lambda),$$

(see Figure 4). The critical curve, which we mention here, indicates a curve that is quickly approached by solutions which then follow this curve on the $H - F$ plane. The curve given by $H = \phi_1(F)$ is a parabola. The threshold we model is the intersection of the two branches of the critical curve, that is the point $F = c/d$, because the hyperparasite population is always near zero below this point and is positive, at least sometimes, above it. Thus, we realise that this results in either positive numbers of hyperparasites, or zero hyperparasites, depending on the fluke population and the value of c/d, which we take now to be equal to 185. The picture shown in Figure 4, can be interpreted either as the phase plane of Eqs. (1) and (2), or as the bifurcation diagram of the fast equation (2), in which the slow variable F is treated as a parameter. Stable and unstable equilibrium points of Eq. (2) are shown by the solid and dashed lines respectively, with the transcritical bifurcation at the intersection point where $F = F_C = c/d$, which is the threshold value ($F_C = 185$).

While a trajectory closely follows one of the curves $H = \phi_k(F)$, the dynamics of the slowly varying fluke population is approximated by the reduced equation

$$\frac{dF}{dt} = a(t) - \delta F - \lambda\phi_k(F). \tag{3}$$

In the vicinity of the critical curve, solutions evolve on timescale equivalent to that of the fluke population, that is slow relative to the hyperparasite population. In our interpretation, the hyperparasite population is detectable when a trajectory is close to the positive branch $H = \phi_1(F)$ of the critical curve, and becomes undetectable when the trajectory switches to the zero branch $H = \phi_0(F) \equiv 0$. The scenario where a trajectory switches from one branch to the other at their intersection point $(F, H) = (F_C, 0)$ is known as an immediate exchange of stabilities, see the descending part of the trajectory in the left panel of Figure 5. It is evident from this figure that a different behaviour also arises where the trajectory stays in the neighbourhood of the unstable zero branch for a certain time after passing the point $F = F_C$, $H = 0$. The trajectory then follows a fast sharp transition from approximately a point $F = F_{th}$, $H = 0$, with $F_{th} > F_C$, to the stable branch $H = \phi_1(F)$, that is the scenario known as a delayed exchange of stability (see the ascending part of the trajectory in the left panel of Figure 5). It is clear therefore, that our system exhibits both *immediate* and *delayed* stability exchange, namely as the populations decrease and increase respectively. The immediate exchange of stability at F_C is defined by the parameters of the system, while the point of delayed stability exchange at F_{th} is defined by the dynamics of the system and varies for different trajectories. In particular, it is dependent on how long the value of $a(t)$ is equal to zero and the minimum value reached by the fluke population, F_{min}.

2.4. Seasonally forced model

We see that, in accordance with the different biologies of the interacting species, the faster timescale of the hyperparasite life cycle results in two thresholds as shown in the left panel of Figure 5. Nevertheless, the transcritical bifurcation point amply illustrates the overall threshold behaviour of the system.

In the above section we developed a model to describe the hyperparasite-fluke interaction within a cockle and the presence of a persistence–type threshold. We now address seasonal effects. Seasonality plays an important role within the system, as effects such as temperature, bird migration, pollution, and other possible abiotic and biotic factors as mentioned earlier, will all lead to variations in the populations under consideration.

As a first approach, we consider Eqs. (1), (2) with a square wave, or switch-type immigration term. This means that the fluke immigration is a positive constant for part of the year, and zero for the rest. This is a realistic model for our system as flukes mature within *S. plana*, and are actively moving and swimming after having been artificially released, only at particular intervals of the year. In particular, Fermer *et al.* (2010) found that these stage cercariae appeared from June through October, and were the dominant developmental stage from July to October. The square wave switching model is also easier to analyse mathematically than other periodic patterns of immigration including the sinusoidal type, to be mentioned later.

We introduce the square wave immigration term $a(t) \equiv a(t+T)$ in Eq. (1) where $a(t) = A > 0$ for part of the year and zero otherwise, such as

$$a(t) = \begin{cases} A, & 0 \leqslant t\,(\text{mod}\ T) \leqslant \tau \\ 0, & \tau < t\,(\text{mod}\ T) < T = 12 \end{cases} \tag{4}$$

where there is immigration over τ months, and the population suffers from natural death and hyperparasitism as before. As an initial approximation, we take the value of τ to be 5, meaning we allow immigration for five months in the year, in keeping with experimental findings described above. A number of important population points and times occur and are shown in the left hand panel of Figure 5.

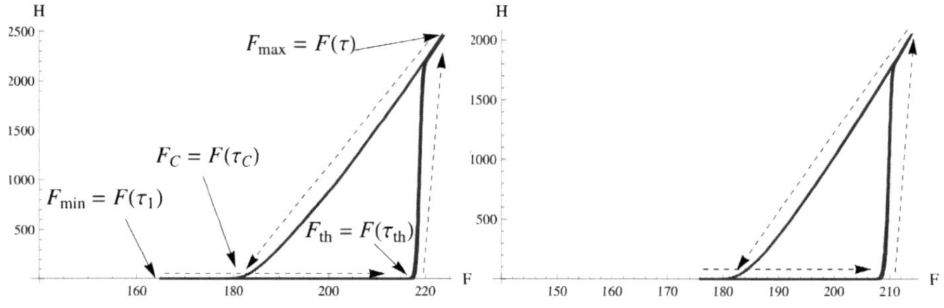

Figure 5. Left: Periodically varying hyperparasite and fluke populations obtained by numerical solution of Eqs. (1) and (2) with $a(t)$ as given by Eq. (4). Parameter values used were as given in Table 1, with $\tau = 5$ in Eq. (4) and $f\lambda/\varepsilon = 5$. Right: Sinusoidally forced system with fluke dynamics given by Eq. (2.7). Here, parameter values used were $a = 10$, $c = 185d$, $d = 1/3000$, $\lambda = 1/300$, $\delta = 1/25$, $\varepsilon = 1/5000$, $f\lambda/\varepsilon = 5$. Quantitative differences between populations under the two immigration scenarios are due to different parameter sets e.g. maximum hyperparasite population is close to 2500 for square wave immigration term and just over 2000 for the sine wave term immigration.

This figure presents the trajectory of a stable T-periodic solution of system (1), (2) and (4). Figure 6 presents the time traces of both of the populations (for clarity, the hyperparasite population is scaled by a constant) over a longer simulation, demonstrating the stability of the system and the presence and absence of the hyperparasite population depending on the oscillation of the fluke population above and below the threshold. In the left hand panel of Figure 5, the fluke population reaches its minimal value F_{min} at the time of the year when the immigration starts, that is at the moment $\tau_1 = 0$ (and then at the same time of the year $t = kT$ every year). At this point, there are no hyperparasites present in the system (more precisely,

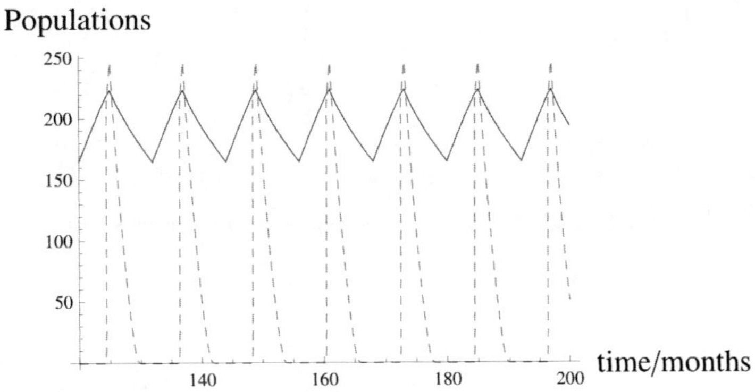

Figure 6. Time trace of the system given by (1), (2) and (4) showing the stable periodic dynamics of the fluke population in blue, hyperparasite population (factor of 1/10) in dashed red, simulated after truncation of the initial variation. Parameters used are as given in Table 1 with $H_0 = F_0 = 100$ at $t = 0$, $\tau = 5$ in Eq. (4) and $f\lambda/\varepsilon = 5$. Yearly periodic behaviour is evident for both populations.

hyperparasites are not detectable as the population H is close to zero and in effect becomes occult). As the fluke population increases with time it eventually reaches a value $F_{th} > F_C$ at a moment τ_{th}, whereby the delayed loss of stability occurs and the system quickly tends to the stable positive branch, $H = \phi_1(F)$ of the critical curve. Further increase in both populations continues, until the instant τ whereupon the fluke population tends to die off through mortality caused by the hyperparasites, together with zero immigration and natural death. The fluke and hyperparasite populations decrease along the positive branch $H = \phi_1(F)$ towards the point $(F, H) = (F_C, 0)$. The hyperparasite population has now collapsed to an exceedingly low level (i.e., becomes undetectable) after the moment τ_C when the fluke population reaches the value F_C. The fluke population continues decreasing to the minimal value F_{min}, which is achieved at the moment T, whereupon the new period begins. As we take monthly time intervals, $T = 12$ in this case.

We see that in the limit of the time separation parameter ε tending to zero, that is, the populations evolve along the critical curve, as indicated in the left panel of Figure 5, the periodic solution is characterised by a few parameters, namely the population values $F_{min} = F(0)$, $F_{th} = F(\tau_{th})$, $F_{max} = F(\tau)$ and the time moments τ_{th}, τ, τ_C. In this limit, the periodic solution follows the zero branch, $H = 0$, between the moments $\tau_1 = 0$ and τ_{th} and the positive parabolic branch, $H = \phi_1(F)$, on the rest of the period, while the transition from the zero branch to the positive branch becomes infinitely fast. It is interesting in this regime to look further at the algebraic consequences due to the ability to integrate the system explicitly.

2.5. Algebraic relations for the limit of the periodic solution

The reduced equation (3) can be integrated explicitly on each branch of the slow manifold. On the time interval $0 \leqslant t \leqslant \tau_{th}$, the reduced equation is obtained by setting $H = 0$ and $a(t) = A$, resulting in a linear equation with the solution

$$F(t) = F_A(t) = \frac{A}{\delta} + e^{-\delta t}\left(F_{min} - \frac{A}{\delta}\right). \tag{5}$$

Journal of Physics: Conference Series **585** (2015) 012013 doi:10.1088/1742-6596/585/1/012013

We can immediately deduce the value for one characteristic population,

$$F_{th} = F_A(\tau_{th}) = \frac{A}{\delta} + e^{-\delta\tau_{th}}\left(F_{min} - \frac{A}{\delta}\right).$$

Between the moments τ_{th} and τ_C, when parasites are present after the threshold time and the solution follows the positive branch $H = \phi_1(F)$, the evolution of the fluke population is governed by the reduced equation:

$$\frac{dF}{dt} = \alpha - \delta F - \frac{F}{f}(-c + dF)$$

where $\alpha = A$ on the time interval $\tau_{th} \leqslant t \leqslant \tau$ and $\alpha = 0$ on the time interval $\tau \leqslant t \leqslant \tau_C$. Separating the variables in this equation and integrating over each of these two time intervals leads to the relations

$$\phi_A(F_{max}) - \phi_A(F_{th}) = \tau - \tau_{th}, \qquad \phi_0\left(c/d\right) - \phi_0(F_{max}) = \tau_C - \tau,$$

with

$$\phi_\alpha(F) = \int \frac{dF}{\mu F^2 + \nu F + \alpha},$$

where $\mu = -d/f$, $\nu = c/f - \delta$. As the discriminant $\nu^2 - 4\,\mu\alpha$ of the denominator of the integrand is positive (we neglect the special case where $c/f = \delta$),

$$\phi_A(F) = \frac{1}{\sigma}\log\left|\frac{2\mu F + \nu - \sigma}{2\mu F + \nu + \sigma}\right|, \qquad \phi_0(F) = \frac{1}{\nu}\log\left|\frac{2\mu F}{2\mu F + 2\nu}\right|,$$

where $\sigma = \sqrt{\nu^2 - 4\mu A}$.

From the moment $t = \tau_C$ to $t = T$ the solution again follows the zero branch $H = 0$ of the slow manifold, like during the time interval $0 \leqslant t \leqslant \tau_{th}$ considered above. However, now $a(t) = 0$ and the reduced equation yields

$$F(t) = F_0(t) = \frac{c}{d}e^{\delta(\tau_C - t)} \tag{6}$$

which gives a result for the minimum fluke population reached, namely,

$$F_{min} = \frac{c}{d}e^{\delta(\tau_C - T)},$$

at the time $t = T$.

Finally, consider the time interval when the parasite population is small, i.e. the trajectory stays near the zero branch of the critical curve from the moment τ_C to the moment $\tau_{th} + T$. The delayed loss of stability phenomenon ensures that, in the limit of vanishing ε, the initial and the final moments of this time interval are related by the integral equation

$$\int_{\tau_C}^{\tau_{th}+T} (-c + F(t)d)\,dt = 0, \tag{7}$$

see for example Butuzov *et al.* (2004). This relation follows from the explicit solution of the linearised Eq. (2) at $H = 0$,

$$\varepsilon\frac{dH}{dt} = (-c + dF)H;$$

the solution reads

$$H(t) = H(t_0)\left(\mathrm{Exp}\left[\frac{1}{\varepsilon}\int_{t_0}^{t}(-c + F(t)\,d)\,dt\right]\right)$$

with a large parameter $1/\varepsilon$ under the exponent. We see that $H(t) = H(t_0)$ if the integral under the exponent is zero, which leads to the relationship (7) between the moments τ_C and $\tau_{th} + T$ in our model.

Using the periodicity of the solution and substituting the expressions (5), (6) for F on the time intervals $0 \le t \le \tau_{th}$, $\tau_C \le t \le T$ in Eq. (7), we obtain

$$c\tau_C - cT - \frac{c}{\delta}e^{\delta(\tau_C - T)} + \frac{c}{\delta} - c\tau_{th} + \frac{dA\tau_{th}}{\delta} + \left(\frac{1}{\delta} - \frac{e^{-\delta\tau_{th}}}{\delta}\right)\left(d\,F_{min} - \frac{dA}{\delta}\right) = 0.$$

In summation, we have obtained five equations with five unknown parameters, namely F_{min}, F_{max}, F_{th}, τ_C, τ_{th}, of the periodic solution, in the limit of small ε:

$$\begin{array}{c} F_{th} = \frac{A}{\delta} + e^{-\delta\tau_{th}}\left(F_{min} - \frac{A}{\delta}\right), \\ \phi_A(F_{max}) - \phi_A(F_{th}) = \tau - \tau_{th}, \\ \phi_0\left(\frac{c}{d}\right) - \phi_0(F_{max}) = \tau_C - \tau, \\ F_{min} = \frac{c}{d}e^{\delta(\tau_C - T)}, \\ c\tau_C - cT - \frac{c}{\delta}e^{\delta(\tau_C - T)} + \frac{c}{\delta} - c\tau_{th} + \frac{dA\tau_{th}}{\delta} + \left(\frac{1}{\delta} - \frac{e^{-\delta\tau_{th}}}{\delta}\right)\left(d\,F_{min} - \frac{dA}{\delta}\right) = 0. \end{array} \qquad (8)$$

2.6. Numerical results

The set of Eqs. (8), can be rewritten to yield two equations with two unknowns, $\psi_1(\tau_{th}, \tau_C) = 0$ and $\psi_2(\tau_{th}, \tau_C) = 0$. These equations are not given explicitly here as their form does not contribute any additional information to understanding of the problem.

Using the known set of parameter values given in Table 1, a plot of both functions gives an estimate for a point of intersection. This estimate is then used to numerically calculate the point of intersection giving the actual values of τ_{th} and τ_C. For the parameter set used in Table 1 and $\tau = 5$ in Eq. (4), the graphical output is shown in Figure 7. The actual solution found using the estimates from the plot yields $\tau_{th} = 4.3$ months and $\tau_C = 9.1$ months.

Using the results for τ_{th} and τ_C, we obtain other parameters from the relational equations. That is, we find that

$$F_{th} = 218,\ F_{min} = 165,\ \text{and}\ F_{max} = 223.$$

Using the algebraic relations to calculate results as above, it is interesting to compare the numerical solutions with the theoretical infinitely fast transition between both branches in the limit $\varepsilon \to 0$. Figure 8 indicates this with the threshold point $F_{th} = 218$ calculated from above. A close correlation between numerics and the limit solution obtained from Eqs. (8) is evident.

2.7. Sinusoidal immigration term

To confirm the applicability of the model, another variation to the immigration of flukes into the cockle is considered, namely that of using a sinusoidal type immigration, as mentioned already. We state this new modification by changing the immigration term in Eq. (1) to

$$\frac{dF}{dt} = a\,(1 + \sin(2\pi t/T)) - \delta F - \lambda H,$$

where the immigration rate is periodic with the period T of one year. Immigration begins again at time $\tau_1 = 0$ where the fluke population is at its minimum F_{min}. The dynamics follow the same pattern as described above, shown in the right hand panel of Figure 5 demonstrating that the qualitative dynamics of the model remain unchanged. However, this type of model does not give us explicit algebraic relations for parameters of the periodic solution in the limit $\varepsilon \to 0$.

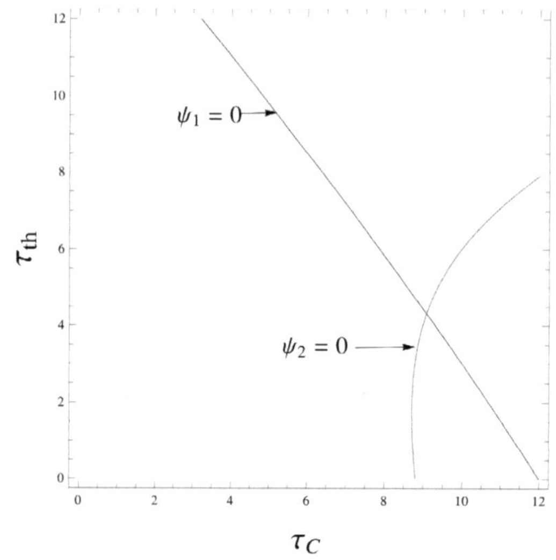

Figure 7. Plot showing intersection of the functions $\psi_1 = 0$ and $\psi_2 = 0$ for estimating τ_{th} and τ_C. Parameters are as given in Table 1, with $\tau = 5$ in Eq. (4) and $f\lambda/\varepsilon = 5$.

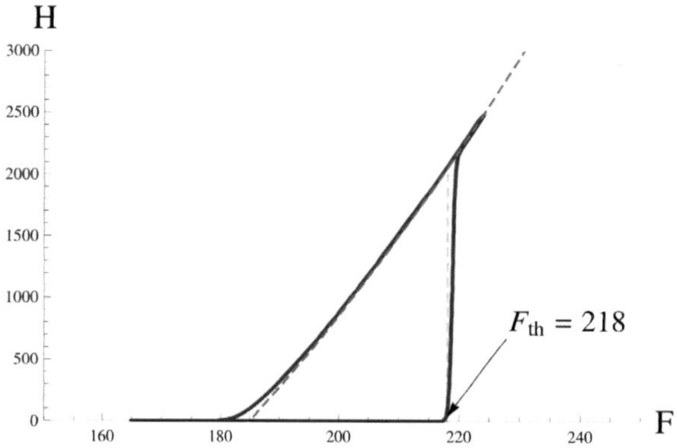

Figure 8. Numerical simulation of the parasite population as a function of fluke population (solid) showing the critical curve and infinitely fast transition (dashed) at the point $F_{th} = 218$ obtained in the limit $\varepsilon \to 0$ from algebraic Eqs. (8) for the parameter set in Table 1.

3. Bird-Shellfish-Parasite model

In the previous section, we have considered a model demonstrating the threshold effect observed for the hyperparasite population to grow and persist in a single cockle, depending on the number of flukes present. We now extend the model to include other species that have an impact on flukes-hyperparasite subsystem, see Figure 1. This can help understanding of the importance that the hyperparasite plays, not just to cockle health, but to the entire ecosystem in bay areas. The periodic immigration terms we used in the flukes equation are artificial, as the actual immigration is derived from the seasonal migration of birds to and from the area. Therefore, we attempt to model the threshold phenomenon using more realistic dynamics related to both the carriers of the initial infection, that is the birds, and the platform for mass reproduction, that is the mollusc *S. plana*. The aim of this section will be to describe each population individually but yet maintain the threshold of an average number of 185 flukes per cockle necessary for the hyperparasite population to grow and persist.

3.1. Oystercatcher population

It can be seen from Figure 1 that the driving force behind the dynamics of the fluke life-cycle is the Oystercatcher (*Haematopus ostralegus*) population. These are the definitive host of the fluke and are the link between the fluke in cockle and those entering *S. plana*, so as to provide a new cycle of immigration and infection. That is, they play the necessary part of closing the life-cycle of the fluke. We refer to these Oystercatchers simply as birds. These birds are also the main predators of the cockle population. Therefore, it is beneficial for the fluke to have birds eat as many cockles as possible, to result in high levels of infection. A break in this point of the chain would mean no flukes present and in turn, no hyperparasites, once the lower limit of the threshold is reached. The fact that the birds consume huge numbers of cockles is itself of interest to the system. Questions as to the lower numbers of cockles needed to entice birds to an area can be asked.

Since these birds are migratory, we introduce a periodic forcing term. We allow the bird population B to follow a basic logistic equation, modified however to be dependent on the cockle population C. This ensures that low numbers of cockles lead to birds having less interest in returning to, or coming to, the area. It is, however, evident that they have an ability to prey on food sources other than that of cockles, but the cockle does provide the main source of food for these birds and thus provides the major stimulus for bringing birds to the area. Thus, the equation for the bird population B will be given by

$$\frac{dB}{dt} = kp(t)CB + k_0B - sB^2,$$

where the migration $p(t)$ is periodic with 12 month cycles; k is a scaling coefficient. It is observed that birds migrate to area's of Ireland in large numbers for Winter months, with peak numbers seen between September and March (Birdwatch Ireland. http://www.birdwatchireland.ie/Default.aspx?tabid=314; accessed 8th March 2012). This forcing term is adjusted to allow for these dates to influence the migratory patterns. As an approximation, we set up the forcing function $p(t)$ so as bird immigration begins on the 10th month, that is October, and all birds leave at the beginning of January; in the interim, there is an added period of growth in bird population, see the left panel of Figure 9.

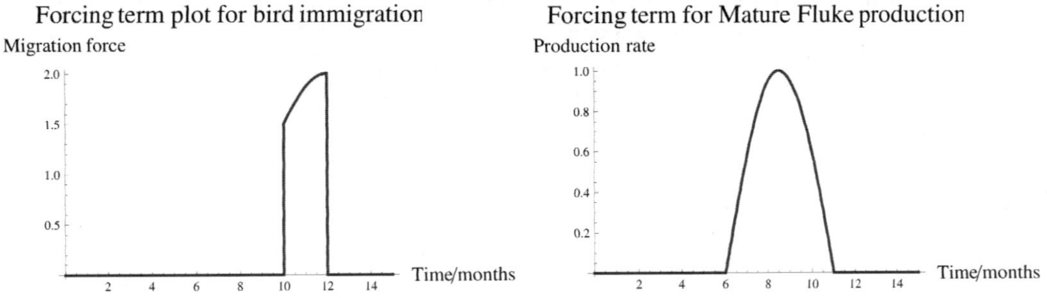

Figure 9. Left: The bird migration forcing term $p(t)$ showing immigration beginning in October, continuing at a declining rate until the end of December, and birds leaving the area in January. This is periodic on a 12 month interval. Right: Plot showing the forcing function $\kappa(t)$ for mature fluke production. Mature flukes were found to be actively swimming and capable of infecting cockles from June (month 6) to November (month 11).

The growth in logistic form due to cockles is accounted for by the CB of the first term and the competition between birds, by the last term. The second term allows for a growth in the bird population even with low cockle numbers indicating alternative food sources. It should be noted that we assume all birds are carriers of the fluke, or at least pick up the fluke infection immediately on returning to an area with infected cockles, and so all of the population can spread flukes to *S. plana*. This assumption is supported by the fact that most areas with fluke infections show cockles with a prevalence of over 80% and reaching 100% (Fermer *et al.*, 2010) and the observation that the birds are each capable of eating in excess of 500 cockles per day.

3.2. Scrobicularia plana *populations*

We refer to *S. plana* just as plana, and we make an initial assumption here that it is reasonable to divide the population into an overall constant population P_0, and the subgroup of fluke-infected plana, $P_I = P_I(t)$. The infected plana population grows due to susceptible plana $P_0 - P_I(t)$ encountering flukes in an infectable stage, that is, those deposited by birds. The population decreases due to a recovery term whereby infected plana recover to carry no flukes. Thus we assume growth in the infected population proportional to the product of the susceptible plana and bird populations, and decline proportional to infected plana, that is standard infection-recovery terms:

$$\frac{dP_I}{dt} = \xi B(P_0 - P_I) - \gamma P_I.$$

According to Fermer *et al.* (2010), the prevalence of fluke infection in plana ranged from over 10% to nearly 25% over the course of a year. The numbers of plana in an area was also found to be much less than those of cockles. Due to the possible high densities of cockles, we assume cockle populations in an area to be of the order of millions. It seems a reasonable assumption that an overall plana population of tens of thousands in an area is acceptable. We use the value of $P_0 = 20000$ and obtain the population of infected plana varying from about 2000 to 5000.

3.3. Fluke populations

The mechanism of flukes progressing through the system is complex and needs to be compartmentalised so as to allow study of individual sections and effects. The model we have chosen is to have a first stage of plana infection, what we call larval flukes, F_{PL}. These are those flukes that initially infect plana from birds. The population will be proportional to the infected plana population ($F_{PL} = \beta P_I$), and we assume small numbers on average per plana. These larval flukes undergo asexual reproduction within the fluke to produce large numbers of mature flukes, F_{PM}, still inhabiting plana. These mature flukes develop only at certain times of the year, so we introduce a periodic forcing term to this equation. The mature flukes go on to infect cockles through emigration from plana and meeting a cockle, but will undergo a death rate either within the plana or during the transmission process, hence we add a natural death rate term to this equation leaving

$$\frac{dF_{PM}}{dt} = r\kappa(t)F_{PL} - \epsilon F_{PM},$$

where the plot of $\kappa(t)$ is shown in right panel of Figure 9; r is a scaling coefficient. These completely developed and freely swimming flukes were observed between June and November, and were the dominant stage from July to October (Fermer *et al.*, 2010).

The next fluke population, F_{CO}, is that proportion of mature flukes that successfully infect a cockle. The number of flukes in cockles should not exceed the number of mature flukes as they have no means of reproducing once inside cockles. The population F_{CO} of flukes in cockles grows proportionally to the product of the mature fluke population and the cockle population, as per typical infection rate term. They have a natural death rate within the cockle and, importantly, suffer due to parasitism by the hyperparasite. Furthermore, we include an additional term in the equation for F_{CO}. As flukes damage cockles, it is reasonable to assume that they in turn suffer loss as this damage increases. We assume that flukes damage cockles as a parasite giving a term of the form $-\omega F_{CO}$ in the cockle equation, yet to be derived. Therefore, flukes themselves will be lost with a compensating term proportional to $-\omega F_{CO} \times (F_{CO}/C)$. We give the full equation as

$$\frac{dF_{CO}}{dt} = f\, C\, F_{PM} - \delta F_{CO} - \nu H - \frac{\zeta \omega F_{CO}^2}{C},$$

where the parameter ζ represents the possibility for over-dispersion as in Eq. (2). Comparing this equation to Eq. (1), the artificial immigration term $a(t)$ is replaced by a dependence on both the cockle population, $C(t)$, and the mature fluke population, $F_{PM}(t)$. The cockle population varies throughout the year due to recruitment and reproduction, as well as predation by birds during Winter months. The second and third terms of both equations remain equivalent. The reason for the additional final term compared to Eq. (1) is because system (1), (2) did not include a measure of cockle damage due to flukes as their population was not relevant to the model.

3.4. Cockle population

The cockle population is assumed to grow logistically in the absence of outside negative effects. It declines due to predation by birds (proportional to the product of cockle and bird populations) and due to fluke damage (proportional to the population of flukes in cockles). The equation is

$$\frac{dC}{dt} = \lambda C - \mu C^2 - \omega F_{CO} - hCB.$$

Parameters ω and h represent the negative parasitic effect of flukes and the predatory effect of birds, respectively.

3.5. Hyperparasite population

The final species we need to model is the hyperparasite. This retains a threshold type behaviour whereby when the population of flukes drops below 185 on average per cockle, the hyperparasite is undetectable. The hyperparasite has a natural death rate, and also a compensating term due its impact on the fluke population, similar to that described earlier for the fluke population and its effect on cockles. The growth rate in this case is reasoned as follows. The hyperparasite population grows following successful transmission to flukes. This occurs within the cockle and so the probability of successful transmission (we can assume that the initial hyperparasite infection remains in a dormant condition prior to infection) will be proportional to an average number of flukes per cockle. Therefore, the probability of successful transmission and growth is proportional to HF_{CO}/C. Again, we assume that hyperparasite population dynamics occur on timescales much less than the other populations, modelled by a small parameter $\varepsilon \ll 1$:

$$\varepsilon \frac{dH}{dt} = -cH + \frac{dHF_{CO}}{C} - \frac{g\nu H^2}{F_{CO}}.$$

Here $c/d = 185$; and g is a coefficient describing over-dispersion. This equation has the same form as Eq. (2), apart from the fact that we now average the recruitment term over the cockle population. This should be reasonable given the amount of homogeneity in a given area.

Journal of Physics: Conference Series **585** (2015) 012013 doi:10.1088/1742-6596/585/1/012013

3.6. Overall system

The system of equations can be summarised as follows:

$$\begin{aligned} F_{PL} &= \beta P_I, \\ \frac{dB}{dt} &= k\left(1 + p\,Cos\left(\frac{2\pi t}{12}\right)\right) CB + k_0 B - sB^2, \\ \frac{dP_I}{dt} &= \xi B\left(P_0 - P_I\right) - \gamma P_I, \\ \frac{dF_{PM}}{dt} &= \kappa(t) F_{PL} - \epsilon F_{PM}, \\ \frac{dC}{dt} &= \lambda C - \mu C^2 - \omega F_{CO} - hCB, \\ \frac{dF_{CO}}{dt} &= f\,C\,F_{PM} - \delta F_{CO} - \nu H - \frac{\zeta \omega F_{CO}^2}{C}, \\ \varepsilon \frac{dH}{dt} &= -cH + \frac{dH F_{CO}}{C} - \frac{g\nu H^2}{F_{CO}}. \end{aligned} \tag{9}$$

It is not easy to provide an estimate for the parameters due to their sheer number and the complexity of the system. However, we wish to be left with reasonable population numbers. We have given hints at some of these numbers above; we will now give a more complete description.

A typical density of cockles can be anything around 1000 per square meter. However, it was found that numbers more typical in the south of Ireland were around 100 - 500 (Fermer *et al.*, 2010). Since the typical area of a bay could be of the magnitude of several acres, it is reasonable to assume that a typical cockle population would be of the order of millions. We will use numbers of around three million cockles in total. *S. plana* numbers are much smaller than those of cockles and we use a fixed population of $P_0 = 20,000$ in the area under consideration. The number of those plana actually infected with flukes ranges from about 10% to 25% giving numbers of at least $2,000$ to $4,000$ infected plana.

Each cockle can be infected with anything from 100 to 3,000 flukes or more (Fermer *et al.*, 2011), meaning the number of flukes in cockles can range from about 10^8 to 10^{10}. These numbers mean that the total number of hyperparasites could be up to the order of 10^{12} or more. The final population, birds, could be assumed at peak times to range in the hundreds to possibly thousands, with essentially absence during warmer Summer months.

3.7. Numerical results

Table 2 presents a typical set of parameters resulting in population numbers within expected ranges for multi-species system (9). In particular, the infected *S. plana* population varies from about $2,000$ to over $2,300$ during a one year cycle. The cockle population has a low of about 2.8 million and a high of about 3.4 million. The average numbers of flukes per cockle is around 200 with some annual variation, and the corresponding hyperparasite population, always present due to the fluke population remaining above the threshold, around $29,000$ per cockle. The bird population dies out, or at least is very small during summer months, and peaks at about 900 during the Winter. The corresponding negative effect that the birds have on cockles can be seen in the dramatic decline in cockle numbers at this time of year. The populations of mature flukes in plana and total numbers of flukes in cockles are also in line with expectations.

3.8. Threshold effect

Figure 10 shows a possibility of the translation of the threshold phenomenon from Eqs. (1), (2) to the more complex system (9). The hyperparasite population is not present for part of the year when the fluke population is lower than the threshold value 185. Once this is reached (or at least, once the exchange of stability occurs) the hyperparasite population grows rapidly, leading to a momentary decline in flukes, before an increase in both populations. There are probably effects of feedback to the system at play here also. The fluke population grows through immigration which is dependent on birds which themselves are dependent on cockles. After this point, there is a steady decline until the fluke population decreases below the threshold and the hyperparasite population goes to zero until the new season of fluke immigration begins. The parameters used are given in Table 3.

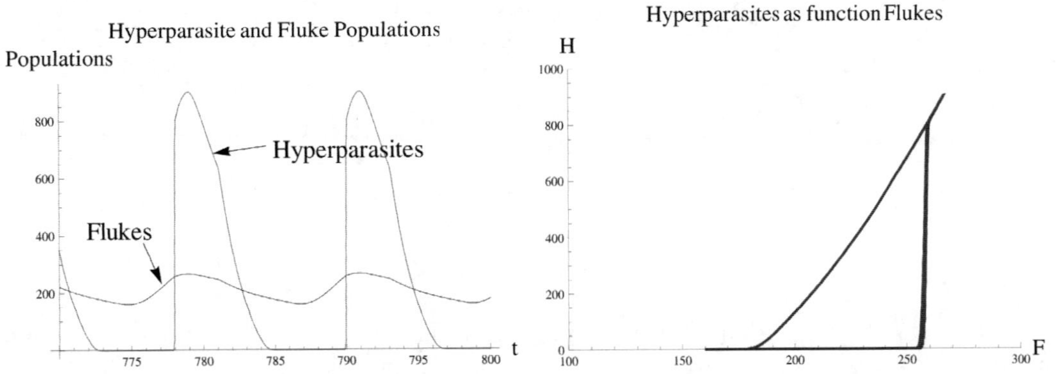

Figure 10. Threshold effect evident for system (9) with parameter values as given in Table 3. Left: Time traces of the fluke and hyperparasite populations. Right: Projection of the periodic solution on the fluke-hyperparasite plane. The hyperparasite can be seen to be absent from the system during periods in which the average number of flukes per cockle drops below 185, that is the threshold type dynamics similar to that demonstrated by system (1), (2), cf. Figure 5.

3.9. The role of the hyperparasite U. legeri *in the ecosystem*
The structure of the interactions of species in Figure 1, and simulations of system (9), suggest that the hyperparasite may play a very important role in the ecosystem. That is, it impacts on the entire system in a much greater way than its relative size would suggest. In order to provide some verification of this hypothesis, we test a scenario where the hyperparasite is effectively removed entirely from the system at some moment in time, and examine the impact that this has on the other populations. The method employed is, using the parameter values as given in Table 2, to solve system (9) for the time $t = 600$ months to ensure that the system reaches the stable periodic regime. From the time $t = 600$ up to the time $t = 1400$, the value of the parameter c is greatly increased. The actual value was increased from $185\,d$ to $30,000\,d$. This ensures that the hyperparasite is present only in extremely small numbers and is essentially wiped out from the system after the moment $t = 600$. Figure 11 shows the effect that this change has on the cockle population. It can be seen that the change is quite dramatic. With the parasite present, the cockle population varies around 3 million, depending on time of year. In the absence of the hyperparasite, this figure drops to some 700,000 cockles, a drop of over 70%. The effects on birds, infected plana and other populations (not shown) are similar. In fact, the bird and infected plana populations drop by almost 80% after the hyperparasite extinction.

The model (9) does not allow for the populations to become zero. Future work may look at a model where it is possible to better quantify the role of the hyperparasite via a bifurcation scenario.

4. Conclusions
We have developed a preliminary model that describes the observed threshold phenomenon in the cockle system. The interaction of this fluke-hyperparasite system appears to be of considerable interest. Gam *et al.* (2008) in a study of a coastal lagoon in Morocco found a higher prevalence of this fluke in older cockles in the inner lagoon compared to the outer lagoon. One suggestion for the difference in prevalence was the higher rate of hyperparasitism in the outer lagoon, with

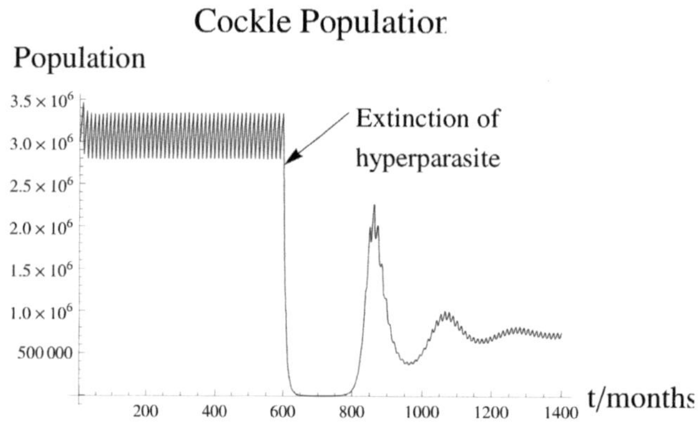

Figure 11. Cockle population in system (9) with parameter values from Table 3. After 600 months, the hyperparasite's death rate is changed to a greatly increased level leading to their extinction. The effect this has on cockles is shown here. A stable periodic solution of over 3 million cockles is reached before the hyperparasite is artificially removed from the system whence there is a dramatic temporary extinction or absence of cockles from the area for almost 200 months. The effects that this could have on an ecosystem may be permanently damaging. However, as the model does not allow for the cockle population to remain completely extinct, after a period of renewal, the cockle population settles to a new periodic pattern varying around the value of 700, 000 cockles. Similar effects are seen in the other populations indicating the importance that the hyperparasite has on the system despite its minute size. This effect is indicative that the hyperparasite is a keystone species.

90% of the trematodes being hyperparasited. It was suggested that this hyperparasite infects and kills *M. minutus* and therefore may control infrapopulations of this trematode (James *et al.* 1977).

Through the use of a relatively standard host-parasite model based on those developed over the past number of decades, we introduced the assumption that the hyperparasite dynamics occur in the system on timescales much faster than that of the fluke. Again this is a classical slow-fast system which arises in engineering, biology and physics. The introduction of the small parameter in the parasite equation gives rise to the threshold, below which no parasites are observed. In fact, due to the delayed loss of stability, two thresholds are actually apparent. With the introduction of a mathematical threshold, we examined the scenario whereby switching of the fluke immigration on and off occurs. This was described to be a realistic occurrence due to the periodic nature of fluke maturation in *S. plana*. It should be noted that it has been observed that although fluke populations can develop in young cockles, parasites do not appear until the cockle is over one year old and there also appears to be the building up of a type of immunity towards flukes in later life. Thus there are varying factors in the immigration rate that may need consideration in a future work. We examined immigration effects and, through a simplification of a switch over a sinusoidal type immigration, we showed that it was possible to derive algebraic relations, which accurately define the periodic solution in the limit where the ratio of the fluke and hyperparasite time scales tends to zero. In particular, we observe both delayed and immediate stability exchange in the hyperparasite population and well-defined indicator times and populations that characterise the periodic dynamics and the thresholds. Assuming parameter values, we used the equations relating the unknowns to derive explicitly the moment of the year when hyperparasite populations start to grow. From this it was possible also to find other points such as the maximum and minimum fluke populations, the threshold fluke population and corresponding times.

Finally, we have considered a seasonally driven multi-species model of an ecosystem shown in Figure 1, which involves bird, and shellfish species typical of a bay or estuarine area, and accounts for typical stages of the parasites' life cycles. We have demonstrated that the persistence threshold effect can be transferred from the cockle-hyperparasite subsystem to the entire ecosystem in this model. We have also provided evidence that the hyperparasite, despite its minute size, can be an important regulator of abundance of co-dependent species in the ecosystem.

Acknowledgments

This work was supported by the European Regional Development Fund (ERDF) through the Ireland Wales Programme (INTERREG 4A). DR acknowledges the support of the Russian Foundation for basic Research through grant 10-01-93112.

References

[1] R M Anderson and R M May 1978 Regulation and stability of host-parasite population interactions: I. Regulatory processes *Journal of Animal Ecology* **47** 219-247

[2] J J Beukema and R Dekker 2005 Decline of recruitment success in cockles and other bivalves in the Wadden Sea: possible role of climate change, predation on postlarvae and fisheries *Mar Ecol Prog Ser* **287** 149-167

[3] E A Bowers *et al.* 1996 The metacercariae of sibling species of Meiogymnophallus, including M. rebecqui comb. nov. (Digenea: Gymnophallidae), and their effects on closely related Cerastoderma host species (Mollusca: Bivalvia) *Parasitology Research* **82** 6 505-510

[4] V F Butuzov *et al.* 2004 Singularly perturbed problems in case of exchange of stabilities *Journal of Mathematical Sciences* **121** 1973-2079

[5] X de Montaudouin, I Kisielewski, G Bachelet and C Desclaux 2000 A census of macroparasites in an intertidal bivalve community, Arcachon Bay, France *Oceanologica Acta* **23** 453-468

[6] X de Montaudouin *et al.* 2009 Digenean trematode species in the cockle Cerastoderma edule: identification key and distribution along the north-eastern Atlantic shoreline *Journal of the Marine Biological Association of the United Kingdom* **89** 543-556

[7] A Deredec and F Courchamp 2003 Extinction thresholds in host-parasite dynamics *Ann Zool Fennici* **40** 115-130

[8] E S Didier, M E Stovall, L C Green, P J Brindley, K Sestak and P J Didier 2004 Epidemiology of microsporidiosis: sources and modes of transmission *Veterinary Parasitology* **126** 145-166

[9] J P Ducrotoy *et al.* 1991 A comparison of the population dynamics of the cockle (Cerastoderma edule, L.) in North-Western Europe *Estuaries and Coasts: Spatial and Temporal Intercomparisons* M Elliott and J-P Ducrotoy (eds) (Olsen & Olsen) 173-184

[10] G W Esch, M A Barger and K J Fellis 2002 The transmission of digenetic trematodes: style, elegance, complexity *Integr Comp Biol* **42** 304312

[11] M Etchechoury and C Muravchik 2003 Nonstandard singular perturbation systems and higher index differential-algebraic systems *Applied Mathematics and Computation* **134** 323-344

[12] J Fermer 2009 Parasitological investigation of softsediment bivalve, with particular reference to the digenean trematode *Meiogymnophallus minutus* Ph D Dissertation, University College Cork, Ireland, October 2009, 143 pages

[13] J Fermer *et al.* 2010 Temporal variation of Meiogymnophallus minutus infections in the first and second intermediate host *Journal of Helminthology* **84** 362368

[14] J Fermer *et al.* 2011 Manipulation of Cerastoderma edule burrowing ability by Meiogymnophallus minutus metacercariae *Journal of the Marine Biological Association of the United Kingdom* **91** 907911

[15] M Gam, H Bazairi, K T Jensen and X de Montaudouin 2008 Metazoan parasites in an intermediate host population near its southern border: the common cockle (Cerastoderma edule) and its trematodes in a Moroccan coastal lagoon (Merja Zerga) *Journal of the Marine Biological Association of the United Kingdom* **88** 357364

[16] M Gam, X de Montaudouin and H Bazairi 2010 Population dynamics and secondary production of the cockle Cerastoderma edule: A comparison between Merja Zerga (Moroccan Atlantic coast) and Arcachon Bay (French Atlantic coast) *Journal of Sea Research* **63** 191-201

[17] W M Getz and J Pickering 1983 Epidemic models: thresholds and population regulation *Am Nat* **121** 6 892-898

[18] R F Hechinger *et al.* 2007 Can parasites be indicators of free-living diversity? Relationships between species richness and the abundance of larval trematodes and of local benthos and fishes Oecologia **151** 82-92
[19] J A P Heesterbeek and M G Roberts 1995 Mathematical models for microparasites of wildlife *Ecology of Infectious Diseases in Natural Populations* B T Grenfell and A P Dobson (eds) (Cambridge: Cambridge University Press) 91-122
[20] R D Holt *et al.* 2003 Impacts of environmental variability in open populations and communities: inflation in sink environments *Theoretical population biology* **64** 315-330
[21] P J Hudson *et al.* 2006 Is a healthy ecosystem one that is rich in parasites? *Trends in Ecology and Evolution* **21** 7 381-385
[22] B L James, A Sannia and E A Bowers 1977 Parasites of birds and shellfish *Problems of a Small Estuary* A Nelson-Smith and E M Bridges (eds) (Swansea: Proc Burry Inlet Symposium **1**) 1-16
[23] P T J Johnson, A Dobson, K D Lafferty, D Marcogliese, J Memmott, S Orlofske, R Poulin and D W Thieltges 2010 When parasites become prey: ecological and epidemiological significance *Trends in Ecology & Evolution* **25** 362-371
[24] L Kalachev, T C Kelly, M J O'Callaghan, A V Pokrovskii and A A Pokrovskiy 2010 Analysis of threshold-type behaviour in mathematical models of the intrusion of a novel macroparasite in a host colony *Math Med Biol* **28** 4 287-333
[25] A M Kuris *et al.* 2008 Ecosystem energetic implications of parasite and free-living biomass in three estuaries *Nature* **454** 515-518
[26] G Lauckner *et al.* 1983 Diseases of mollusca: Bivalvia *Diseases of Marine Animals Vol. 2 Introduction, Bivalvia to Scaphopoda* O Kinne (ed) (Hamburg: Biologische Anstalt Helgoland) 477-961
[27] K N Mouritsen and R Poulin 2002 Parasitism, community structure and biodiversity in intertidal ecosystems *Parasitology* **124** S101-S117
[28] M Nizette, D Rachinskii, A Vladimirov and M Wolfrum 2006 Pulse interaction via gain and loss dynamics in passive mode locking *Physica D: Nonlinear Phenomena* **218** 1 95-104
[29] T R Raffel *et al.* 2008 Parasites as predators: unifying natural enemy ecology *Trends in Ecology and Evolution* **23** 11 610-618
[30] K Rohde 1994 Niche restriction in parasites: proximate and ultimate causes *Parasitology* **109** S69-S84
[31] F Russell-Pinto 1990 Differences in infestation intensity and prevalence of hinge and mantle margin Meiogymnophallus minutus Metacercariae (Gymnophallidae) in Cerastoderma edule (Bivalvia): Possible species coexistence in Ria de Aveiro *J Parasitology* **76** 5 653-659
[32] F Russell-Pinto *et al.* 2006 Digenean Larvae parasitizing Cerastoderma Edule (Bivalvia) and Nassarius Reticulatus (Gastropoda) from Ria de Aveiro, Portugal *J Parasitology* **92** 2 319-332
[33] D W Thieltges and K Reise 2006 Metazoan parasites in intertidal cockles Cerastoderma edule from the northern Wadden Sea *Journal of Sea Research* **56** 284-293
[34] E R Troemel *et al.* 2008 Microsporidia Are Natural Intracellular Parasites of the Nematode Caenorhabditis elegans *PLoS Biol* **6** 12 e309
[35] V Veliov 1997 A generalization of the Tikhonov Theorem for singularly perturbed differential inclusions *Journal of dynamical and control systems* **3** 291-319
[36] A M Wegeberg and K T Jensen 1999 Reduced survivorship of Himasthla (Trematoda, Digenea)-infected cockles (Cerastoderma edule) exposed to oxygen depletion *Journal of Sea Research* **42** 4 325-331
[37] A M Wegeberg and K T Jensen 2003 In situ growth of juvenile cockles, Cerastoderma edule, experimentally infected with larval trematodes (Himasthla interrupta) *Journal of Sea Research* **50** 1 37-43

Parameter	Value
β	10
k	8×10^{-6}
p	$0 \leq p(t) \leq 2$
k_0	0.088
s	0.054
ξ	8×10^{-6}
P_0	$20,000$
γ	0.014
r	$50,000$
κ	$0 \leq \kappa(t) \leq 1$
ϵ	0.06
λ	0.15812
μ	2×10^{-8}
ω	0.0004
h	0.0001
f	3.33×10^{-7}
δ	0.1
ν	0.05
ζ	8
ε	0.001
c	$185d$
d	0.0667
g	0.03

Table 2. Parameters used for numerical simulations of system (9).

Parameter	Value
β	10
k	10^{-6}
p	$0 \leq p(t) \leq 2$
k_0	0.88
s	0.054
ξ	10^{-5}
P_0	$20,000$
γ	0.01
r	$25,000$
κ	$0 \leq \kappa(t) \leq 1$
ϵ	1
λ	0.15
μ	2×10^{-8}
ω	0.00025
h	0.00009
f	1.25×10^{-7}
δ	0.0001
ν	0.0256
ζ	0.00025
ε	0.0001
c	$185d$
d	0.0014
g	1.3

Table 3. Parameters used for numerical simulations of system (9).

Journal of Physics: Conference Series **585** (2015) 012014 doi:10.1088/1742-6596/585/1/012014

New approaches to the modelling of multi-component fuel droplet heating and evaporation

Sergei S Sazhin[1], **Ahmed E Elwardany**[1&], **Morgan R Heikal**[1*]

[1]Sir Harry Ricardo Laboratories, Centre for Automotive Engineering, School of Computing, Engineering and Mathematics, Faculty of Science and Engineering, University of Brighton, Brighton, BN2 4GJ, UK. S.Sazhin@brighton.ac.uk

[&]Current address: King Abdullah University of Science and Technology 4700 Thuwal 23955-6900, Saudi Arabia

[*]Current address: Department of Mechanical Engineering, Universiti of Teknologies PETRONAS, Bandar Sri Iskandar 31750 Tronoh, Perak Darul Ridzuan, Malaysia

E-mail: `S.Sazhin@brighton.ac.uk`

Abstract. The previously suggested quasi-discrete model for heating and evaporation of complex multi-component hydrocarbon fuel droplets is described. The dependence of density, viscosity, heat capacity and thermal conductivity of liquid components on carbon numbers n and temperatures is taken into account. The effects of temperature gradient and quasi-component diffusion inside droplets are taken into account. The analysis is based on the Effective Thermal Conductivity/Effective Diffusivity (ETC/ED) model. This model is applied to the analysis of Diesel and gasoline fuel droplet heating and evaporation. The components with relatively close n are replaced by quasi-components with properties calculated as average properties of the *a priori* defined groups of actual components. Thus the analysis of the heating and evaporation of droplets consisting of many components is replaced with the analysis of the heating and evaporation of droplets consisting of relatively few quasi-components. It is demonstrated that for Diesel and gasoline fuel droplets the predictions of the model based on five quasi-components are almost indistinguishable from the predictions of the model based on twenty quasi-components for Diesel fuel droplets and are very close to the predictions of the model based on thirteen quasi-components for gasoline fuel droplets. It is recommended that in the cases of both Diesel and gasoline spray combustion modelling, the analysis of droplet heating and evaporation is based on as little as five quasi-components.

1. Introduction

In a number of our earlier papers it was demonstrated that the choice of the model for droplet heating and evaporation leads to a noticeable effect on the predicted autoignition delay in Diesel fuel sprays [1, 2]. It was pointed out that a widely used model for droplet heating and evaporation, based on the assumption that the liquid thermal conductivity is infinitely large, leads to considerable overestimation of the autoignition delay, especially for high gas temperatures. It was recommended that the effect of the finite value of this conductivity be taken into account in spray autoignition modelling using the Effective Thermal Conductivity (ETC) model [3].

One of the main limitations of the analysis presented in [1, 2] is that it was based on the assumption that Diesel fuel can be approximated as a mono-component fuel. This assumption

Journal of Physics: Conference Series **585** (2015) 012014 doi:10.1088/1742-6596/585/1/012014

is obviously too crude for realistic Diesel fuels which consist of hundreds of various components. The models suggested in the literature to deal with such complex fuels are based on the probabilistic analysis of a large number of components (e.g. Continuous Thermodynamics approach [4]-[11], the Distillation Curve Model [12], the combination of the Continuous Thermodynamics approach and the Discrete Component Model [13], and the Quadrature Method of Moments [14, 15]). All these models, however, are based on a number of additional simplifying assumptions, including the assumption that species inside droplets mix infinitely quickly or do not mix at all. As follows from our analysis of heating and evaporation of bi-component droplets [16, 17], this assumption can lead to large errors.

A new approach to the modelling of heating and evaporation of multi-component droplets, suitable for the case when a large number of components is present in the droplets, is suggested in [18, 19]. This approach is based on the introduction of quasi-components, which are treated as hypothetical components, but the number of these quasi-components is much smaller than the number of actual components. In contrast to the previously suggested models, designed for large numbers of components, the new model takes into account the diffusion of liquid species and thermal diffusion. This model was called the 'quasi-discrete model'.

Despite a number of limitations, we believe that it is the most advanced approach when compared with others suggested for the modelling of heating and evaporation in complex hydrocarbon sprays, including Diesel and gasoline fuel sprays. The modelling of heating and evaporation of these sprays is integral to the modelling of their autoignition.

The aim of this paper is to present a brief summary of the essence of this model and show some new results.

2. Quasi-discrete model

As in the case of the Continuous Thermodynamics approach, the quasi-discrete model is based on the introduction of the distribution function $f_m(I)$ such that:

$$\int_{I_1}^{I_2} f_m(I)\mathrm{d}I = 1, \tag{1}$$

where I is the property of the component, f_m characterises the relative contribution of the components having this property in the vicinity of I, I_1 and I_2 are limiting values of this property.

Although molar mass M is almost universally used to describe the property I, this choice is certainly far from being a unique one. Remembering that most practically important hydrocarbon fuels consist mainly of molecules of the type C_nH_{2n+2} (alkanes), where $n \geq 5$ for liquid fuels, it is more practical to write the distribution function f_m as a function of the carbon number n rather than M [9]. Note that

$$M = 14n + 2, \tag{2}$$

where M is measured in kg/kmole.

As in [18] we assume that $f_m(n)$ can be approximated as:

$$f_m(n) = C_m(n_0, n_f)\frac{(M(n)-\gamma)^{\alpha-1}}{\beta^\alpha\Gamma(\alpha)}\exp\left[-\left(\frac{M(n)-\gamma}{\beta}\right)\right], \tag{3}$$

where $n_0 \leq n \leq n_f$, subscripts $_0$ and $_f$ stand for initial and final, $\Gamma(\alpha)$ is the Gamma function, α and β are parameters that determine the shape of the distribution, γ determines the original shift,

$$C_m(n_0, n_f) = \left\{\int_{n_0}^{n_f}\frac{(M(n)-\gamma)^{\alpha-1}}{\beta^\alpha\Gamma(\alpha)}\exp\left[-\left(\frac{M(n)-\gamma}{\beta}\right)\right]\mathrm{d}n\right\}^{-1}. \tag{4}$$

This choice of C_m assures that $\int_{n0}^{n_f} f_m(n)\mathrm{d}n = 1$.

As in [18] we use the following approximation for the dependence of the saturation vapour pressure (in MPa) on n:

$$p^{\mathrm{sat}}(n) = \exp\left(A(n) - \frac{B(n)}{T - C(n)}\right), \tag{5}$$

where T is in K,

$$A(n) = 6.318\, n^{0.05091}, \quad B(n) = 1178\, n^{0.4652}, \quad C(n) = 9.467\, n^{0.9143}.$$

From the Clausius-Clapeyron equation it follows that [9]

$$L = -\frac{R_u}{M(n)} \frac{\mathrm{d} \ln p^{\mathrm{sat}}(n)}{\mathrm{d}(1/T)}, \tag{6}$$

where R_u is the universal gas constant. Remembering (5), Formula (6) can be rewritten as:

$$L = \frac{R_u B(n) T^2}{M(n)(T - C(n))^2}. \tag{7}$$

Following [18], this formula will be used in our analysis.

In [18] the dependence of liquid density, viscosity, specific heat capacity and thermal conductivity on n was ignored. In our analysis this dependence is taken into account, following [19], based on published data for liquid density [20], viscosity [21], specific heat capacity [22] and thermal conductivity [23] (see Figs. A1-A4 of [19]).

As follows from (5), (7) and the results for liquid density, viscosity, specific heat capacity and thermal conductivity, the transport and thermodynamic properties of the fuel components are relatively weak functions of n. In this case, following [18, 19], it would be sensible to assume that the properties of hydrocarbons in a certain narrow range of n are close, and replace the continuous distribution (3) with a discrete one, consisting of N_f quasi-components with carbon numbers

$$\overline{n}_j = \frac{\int_{n_{j-1}}^{n_j} n f_m(n)\mathrm{d}n}{\int_{n_{j-1}}^{n_j} f_m(n)\mathrm{d}n}, \tag{8}$$

the corresponding molar fractions

$$X_j = \int_{n_{j-1}}^{n_j} f_m(n)\mathrm{d}n, \tag{9}$$

and mass fractions

$$Y_j = \frac{M(\overline{n}_j) X_j}{\sum_{j=1}^{j=N_f} [M(\overline{n}_j) X_j]}, \tag{10}$$

where j is an integer in the range $1 \le j \le N_f$. We assume that all $n_j - n_{j-1}$ are equal, i.e. all quasi-components have the same range of values of n. For the case when $N_f = 1$ this approach reduces the analysis of multi-component droplets to mono-component ones.

These new quasi-components are not the actual physical hydrocarbon components ($\overline{n}_j$ are not integers in the general case). Hence we called this model a quasi-discrete model. These quasi-components are treated as actual components in the conventional Discrete Component Model, including taking into account diffusion of liquid species in droplets. This model is expected to be particularly useful when N_f is much less than the number of actual species in the hydrocarbon mixture.

Partial pressures of individual quasi-components can be estimated as (Raoult's law is assumed to be valid):

$$p_v(\overline{n}_j) = X_{lsi}(\overline{n}_j)p^{\mathrm{sat}}(\overline{n}_j), \tag{11}$$

where X_{lsi} is the molar fraction of liquid quasi-components at the surface of the droplet, $p^{\mathrm{sat}}(\overline{n}_j)$ is determined by Eq. (5).

Having replaced n with $\overline{n}_j$ we obtain the required values of p^{sat}, L, liquid density, viscosity, specific heat capacity and thermal conductivity for all quasi-components.

Note that, strictly speaking, the viscosity and thermal conductivity of the mixture are not equal to those for $\overline{n}_j$ (see [24] for details). The difference between them, however, is expected to be relatively small and will be ignored in our analysis.

3. The diffusion of quasi-components

Equations for mass fractions of quasi-components $Y_j \equiv Y_j(t, R)$ inside droplets are used in the following form:

$$\frac{\partial Y_j}{\partial t} = D_l \left(\frac{\partial^2 Y_j}{\partial R^2} + \frac{2}{R}\frac{\partial Y_j}{\partial R} \right), \tag{12}$$

where $j > 1$, D_l is the liquid mass diffusivity. Eq. (12) is solved with the following boundary condition [16, 17]:

$$\alpha(\epsilon_j - Y_{js}) = -D_l \left. \frac{\partial Y_j}{\partial R} \right|_{R=R_d-0}, \tag{13}$$

and the initial condition $Y_j(t=0) = Y_{j0}(R)$, where $Y_{js} = Y_{js}(t)$ are liquid components' mass fractions at the droplet's surface,

$$\alpha = \frac{|\dot{m}_d|}{4\pi\rho_l R_d^2}, \tag{14}$$

R_d is the droplet radius, $\dot{m}_d$ is the rate of mass loss by droplets, ρ_l is the total liquid density.

Assuming that species concentrations in the ambient gas are equal to zero ($Y_{v\infty} = 0$), the values of ϵ_j can be found from the following relation [16]:

$$\epsilon_j = \frac{Y_{vjs}}{\sum_j Y_{vjs}}, \tag{15}$$

where the subscript $_v$ indicates the vapour phase. We assume that ϵ_j are still defined by Eq. (15) even in the case when these concentrations are not equal to zero, which is expected in the case of a coupled solution. In contrast to a widely used approach, based on the numerical solution to Eq. (12), our analysis is based on its analytical solution [16, 17]:

$$\begin{aligned} Y_{li} = \epsilon_i + \frac{1}{R}\Bigg\{ &\left[\exp\left[D_l \left(\frac{\lambda_0}{R_d} \right)^2 t \right] [q_{i0} - Q_0\epsilon_i] \sinh\left(\lambda_0 \frac{R}{R_d} \right) \right. \\ &+ \left. \sum_{n=1}^{\infty} \left[\exp\left[-D_l \left(\frac{\lambda_n}{R_d} \right)^2 t \right] [q_{in} - Q_n\epsilon_i] \sin\left(\lambda_n \frac{R}{R_d} \right) \right] \right\}, \end{aligned} \tag{16}$$

where λ_0 and λ_n ($n \geq 1$) are solutions to equations

$$\tanh\lambda = -\frac{\lambda}{h_0} \quad \text{and} \quad \tan\lambda = -\frac{\lambda}{h_0}.$$

respectively, $h_0 = -\left(1 + \frac{\alpha R_d}{D_l}\right)$,

$$Q_n = \begin{cases} -\frac{1}{||v_0||^2}\left(\frac{R_d}{\lambda_0}\right)^2 (1+h_0)\sinh\lambda_0 & \text{when} \quad n=0 \\ \frac{1}{||v_n||^2}\left(\frac{R_d}{\lambda_n}\right)^2 (1+h_0)\sin\lambda_n & \text{when} \quad n \geq 1 \end{cases} \tag{17}$$

$$q_{in} = \frac{1}{||v_n||^2}\int_0^{R_d} RY_{li0}(R)v_n(R)\mathrm{d}R, \quad n \geq 0, \tag{18}$$

$$v_0(R) = \sinh\left(\lambda_0 \frac{R}{R_d}\right), \quad v_n(R) = \sin\left(\lambda_n \frac{R}{R_d}\right), \quad n \geq 1.$$

The average mass fraction of species in a moving droplet can still be correctly predicted by Equation (12), with appropriate boundary conditions, if the liquid diffusivity D_l is replaced with the so-called effective diffusivity D_{eff}

$$D_{\text{eff}} = \chi_Y D_l, \tag{19}$$

where the coefficient χ_Y varies from 1 to 2.72 and can be approximated as:

$$\chi_Y = 1.86 + 0.86\tanh\left[2.225\log_{10}\left(\mathrm{Re}_l\mathrm{Sc}_l/30\right)\right], \tag{20}$$

Re_l is the Reynolds number, based on liquid transport properties and the maximum surface velocity inside droplets, Sc_l is the liquid Schmidt number.

The analytical solution (16) is incorporated into the numerical scheme when the effect of droplet evaporation is taken into account. Recirculation inside droplets is taken into account based on the Effective Thermal Conductivity/Effective Diffusivity (ETC/ED) model [16, 17]. The solution described in [16, 17] has been generalised in [25] to take into account the effect of the moving droplet boundary due to the evaporation process.

4. Application to Diesel and gasoline fuel droplets

Following [9] we assume the following values of parameters for the distribution function (3) for Diesel and gasoline fuels.

Fuel	α	β (kg/kmole)	γ (kg/kmole)	n_0	n_f
Diesel	18.5	10	0	5	25
Gasoline	5.7	15	0	5	18

Table 1

As in [18, 19], we assumed that the initial Diesel fuel droplet temperature is equal to 300 K, and is homogeneous throughout its volume. Gas temperature and pressure are assumed to be equal to 880 K and 3 MPa respectively. The initial composition of droplets is described by distribution function (3) with the values of parameters given in Table 1.

Plots of droplet surface temperature T_s and radius R_d versus time for the initial radius equal to 10 μm and velocity 1 m/s are shown in Fig. 1. The droplet velocity is assumed to be constant during the whole process. The calculations were performed for the case of $N_f = 1$ (one quasi-component droplet, $\overline{n} = 12.56$), $N_f = 5$ (five quasi-component droplet) and $N_f = 20$ (twenty quasi-component droplet), using the ETC/ED model. The latter case is expected to reproduce adequately the properties of actual Diesel fuels [9].

As one can see from this figure, the droplet radii R_d and surface temperature T_s, predicted by the ETC/ED model, using one and twenty quasi-components are noticeably different,

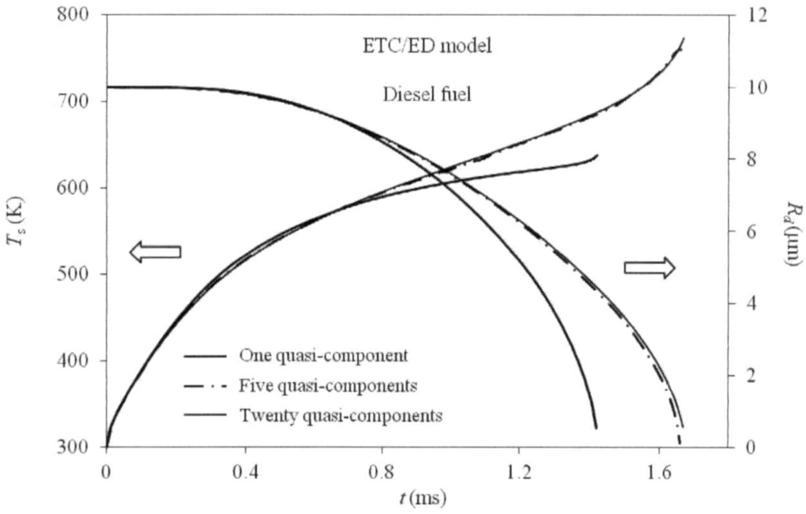

Figure 1. Plots of T_s and R_d, predicted by the ETC/ED model, versus time for 1, 5 and 20 quasi-components. The initial droplet radius and temperature are assumed to be equal to 10 μm and 300 K respectively; the droplet velocity is assumed to be equal to 1 m/s and its changes during the heating and evaporation process are ignored; gas temperature is assumed equal to 880 K.

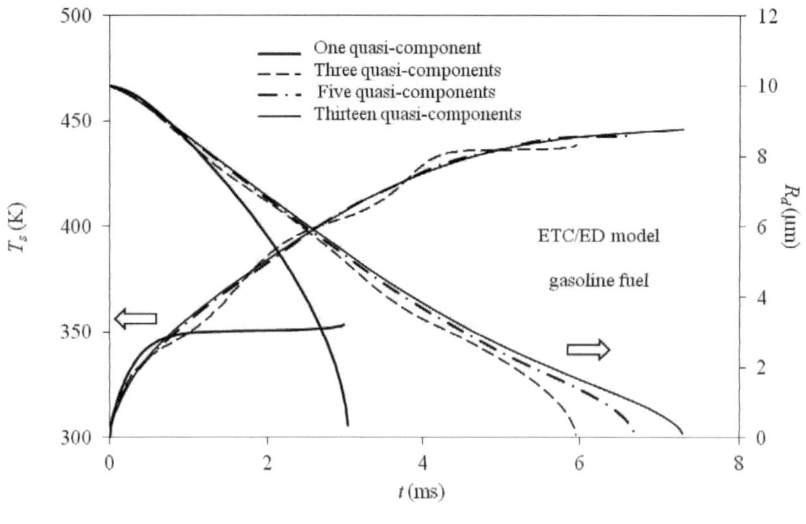

Figure 2. The same as Fig. 1 but for gasoline fuel droplets with velocity equal to 10 m/s, gas temperature equal to 450 K and pressure equal to 0.3 MPa.

especially at the final stages of droplet heating and evaporation. The model, using twenty quasi-components predicts higher surface temperatures and longer evaporation time compared with the model using one quasi-component. This can be related to the fact that at the final stages of droplet evaporation the species with large n become the dominant. These species evaporate more slowly than the species with lower n and have higher wet bulb temperatures. At the same time, the predictions of the model using just five quasi-components are almost identical with the predictions of the model using twenty quasi-components, in agreement with the earlier results reported in [18].

Plots similar to those shown in Fig. 1, but for gasoline fuel droplets in realistic conditions for gasoline engines, are presented in Fig. 2. Following [26], we assume that gas temperature is equal to 450 K, gas pressure is equal to 0.3 MPa and droplet velocity is equal to 10 m/s. As in the case shown in Fig. 1, we assumed that the initial droplet temperature is equal to 300 K, and is homogeneous throughout its volume, while the droplet initial radius is equal to 10 μm.

The calculations were performed for the case of $N_f = 1$, $N_f = 3$, $N_f = 5$ and $N_f = 13$, using the ETC/ED model. The latter case is expected to reproduce adequately the properties of actual gasoline fuels [9].

Comparing Figs. 1 and 2, one can see that in the latter case the difference between the predicted temperatures and droplet radii for one and thirteen quasi-components is much more visible than in the former. This can be attributed to much slower evaporation for the case shown in Fig. 2, compared with the case shown in Fig. 1. At the same time, the predictions of the model using just three and especially five quasi-components are reasonably close to the predictions of the model using thirteen quasi-components. The general trends of the curves shown in Fig. 2 are similar to the ones shown in Fig. 1. In the case when thirteen quasi-components are considered, at the end of the evaporation process, mainly the heavier components in the droplets remain. These can reach higher temperatures and evaporate more slowly compared with the light and middle-range components.

Note that the analysis described above is based on the hydrodynamic model for droplet evaporation. The kinetic and molecular dynamic effects will be taken into account during further developments of this model. Some of the most recent results in this direction are described in [27, 28].

5. Conclusions

The previously suggested quasi-discrete model for heating and evaporation of complex multi-component hydrocarbon fuel droplets is described. The dependence of density, viscosity, heat capacity and thermal conductivity of liquid components on carbon numbers and temperatures is taken into account. This model is applied to the modelling of Diesel and gasoline fuel droplet heating and evaporation.

The model is based upon the assumption that properties of the components vary relatively slowly from one component to another and depend on a single parameter: the number of carbon atoms in the components (carbon number: n). The components with relatively close n are replaced with quasi-components with properties calculated as average properties of the *a priori* defined groups of actual components. Thus the analysis of the heating and evaporation of droplets consisting of many components is replaced with the analysis of the heating and evaporation of droplets consisting of relatively few quasi-components. The effects of temperature gradient and quasi-component diffusion inside droplets are taken into account. The analysis is based on the Effective Thermal Conductivity/Effective Diffusivity (ETC/ED) model.

It is demonstrated that for Diesel fuel droplets the predictions of the model based on five quasi-components are almost indistinguishable from the predictions of the model based on twenty quasi-components (maximal number of quasi-components). These are noticeably different from the predictions of the model, based on one quasi-component, widely used in spray combustion modelling. A similar conclusion was drawn for gasoline fuel droplets with the maximal number of quasi-components equal to thirteen. It is recommended that in the cases of both Diesel and gasoline spray combustion modelling, the analysis of droplet heating and evaporation is based on five quasi-components.

Journal of Physics: Conference Series **585** (2015) 012014 doi:10.1088/1742-6596/585/1/012014

Acknowledgments

The authors are grateful to the European Regional Development Fund Franco-British INTERREG IVA (Project C5, Reference 4005) and EPSRC (grant EP/H001603/1) for financial support of the work on this project.

References

[1] Sazhin S S, Abdelghaffar W A, Sazhina E M, Heikal M R 2005 *Int. J Thermal Science* **44** 610.

[2] Sazhin S S, Martynov S B, Kristyadi T, Crua C and Heikal M R 2008 *Int. J of Engineering Systems Modelling and Simulation* **1** 1.

[3] Sazhin S S 2006 *Prog. in Energy and Combustion Science* **32** 162.

[4] Tamim J and Hallett W L H 1995 *Chem Eng. Sci.* **50** 2933.

[5] Lippert A M and Reitz R D 1997 SAE Technical Paper 972882.

[6] Hallett W L H 2000 *Combust. Flame* **121** 334.

[7] Zhu G-S and Reitz R D 2002 *Int. J Heat and Mass Transfer* **45** 495.

[8] Pagel S, Stiesch G and Merker G P 2002 Modelling of evaporation of a multicomponent fuel. In Taine J, editor. Proceedings of the Twelfth International Heat Transfer Conference, V. 1. Grenoble (August 18 - 23, 2002), Paris: Editions Scientifique et Medicale Elsevier SAS; 2002, CD ROM.

[9] Arias-Zugasti M and Rosner D E 2003 *Combust. Flame* **135** 271.

[10] Abdel-Qader Z and Hallett W L H 2005 *Chem. Eng. Sci.* **60** 1629.

[11] Rivard E and Brüggemann D 2010 *Chem. Eng. Sci.* **65** 5137.

[12] Burger M, Schmehl R, Prommersberger K, Schäfer O, Koch R and Wittig S 2003 *Int. J Heat and Mass Transfer* **46** 4403.

[13] Zhang L and Kong S-C 2010 *Combust. Flame* **157** 2165.

[14] Laurent C 2008 Dévelopment et validation de modèles d'évaporation multi-composant. Thèse de l'Institut Supériour de l'Aéronautique et de l'Espace de Toulouse.

[15] Laurent C, Lavergne G and Villedieu P 2009 *C.R. Mechanique* **337** 449.

[16] Sazhin S S, Elwardany A, Krutitskii P A, Castanet G, Lemoine F, Sazhina E M and Heikal M R 2010 *Int. J of Heat and Mass Transfer* **53** 4495.

[17] Sazhin S S, Elwardany A, Krutitskii P A, Deprédurand V, Castanet G, Lemoine F, Sazhina E M and Heikal M R 2011 *Int. J of Thermal Science* **50** 1164.

[18] Sazhin S S, Elwardany A, Sazhina E M and Heikal M R 2011 *Int. J of Heat and Mass Transfer* **54** 4325.

[19] Elwardany A and Sazhin S S 2012 *Fuel* **97** 685.

[20] Yaws C L(Editor) 2008 Thermophysical Properties of Chemicals and Hydrocarbons. William Andrew Inc.

[21] Mehrotra A K 1994 *The Canadian Journal of Chemical Engineering* **72** 554.

[22] van Miltenburg J C 2000 *Thermochimica Acta* **343** 57.

[23] Yaws C L 1995 Handbook of Thermal Conductivity. Vol. 2 (Organic compounds, C_5 to C_7) and Vol. 3 (Organic compounds, C_8 to C_{28}). Gulf Publishing Company. Houston, London, Paris, Zurich, Tokyo.

[24] Sazhin S S, Kristyadi T, Abdelghaffar W A and Heikal M R 2006 *Fuel* **85** 1613.

[25] Gusev I G, Krutitskii P A, Sazhin S S, Elwardany A 2012 *Int. J of Heat and Mass Transfer* **55** 2014.

[26] Basshuysen R V 2009 Gasoline Engine with Direct Injection: Processes, Systems, Development, Potential, 1st Edition, GWV Fachverlage GmbH, Wiesbaden (978-8348-0670-3).

[27] Xie J-F, Sazhin S S and Cao B-Y 2011 *Physics of Fluids* **23** 112104; doi:10.1063/1.3662004.

[28] Shishkova I N, Sazhina S S, Xie J-F 2012 *J Computational Physics of Fluids* (in press).

Journal of Physics: Conference Series **585** (2015) 012015 doi:10.1088/1742-6596/585/1/012015

Fact and fictions in FX arbitrage processes

Rod Cross[1] **and Victor Kozyakin**[2]

[1]Department of Economics, University of Strathclyde, Sir William Duncan Building, 180 Rottenrow, Glasgow G4 0GE, Scotland.

[2]Institute of Information Transmission Problems, Russian Academy of Sciences, Bolshoi Karetny lane 19, Moscow 127994 GSP-4, Russia.

E-mail: rod.cross@strath.ac.uk, kozyakin@iitp.ru

Abstract. The efficient markets hypothesis implies that arbitrage opportunities in markets such as those for foreign exchange (FX) would be, at most, short-lived. The present paper surveys the fragmented nature of FX markets, revealing that information in these markets is also likely to be fragmented. The "quant" workforce in the hedge fund featured in *The Fear Index* novel by Robert Harris would have little or no reason for their existence in an EMH world. The four currency combinatorial analysis of arbitrage sequences contained in [1] is then considered. Their results suggest that arbitrage processes, rather than being self-extinguishing, tend to be periodic in nature. This helps explain the fact that arbitrage dealing tends to be endemic in FX markets.

1. Introduction

An arbitrage operation involves taking advantage of discrepancies in the prices at which a particular good or asset can be bought or sold, such operations being distinct from speculation in that they involve minimal risk or capital exposure. In the economics and finance literature the "law of one price" postulates that the exploitation of opportunities to profit from arbitrage, if buying prices are lower than selling prices, will lead goods or assets that are in some sense "identical" to have the same price.

The present paper focuses on arbitrage operations in foreign exchange (FX) markets. The Bank for International Settlements [2, p. 5] offers a useful distinction between different arbitrage strategies in FX markets. *Classical arbitrage* exploits differences between quoted exchange rates and the exchange rates implied by no-arbitrage conditions, such as that cross-exchange rates are consistent with the exchange rates for currency pairs. *Latency arbitrage* takes advantage of time lags between market-moving trades being initiated and market-makers updating their FX price quotes. *Liquidity imbalance strategies* aim to detect order book imbalances for currency pairs and pricing discrepancies between different trading platforms. *Complex event* processing aims to exploit properties of foreign exchange rates such as momentum, mean-reversion, correlation with other exchange rates or reaction to news releases. The present paper focuses on *classical arbitrage strategies*.

Section 2 of this paper considers the institutional context in which FX arbitrage strategies can be executed. In contrast to the textbook fiction of a single FX market, there are a variety of FX trading platforms distinguished not only by geographical location and time zone but also by the means by which transactions are executed. Electronic FX trading platforms have grown in importance in the last decade, but a substantial proportion of transactions are still conducted by banks dealing directly with end-user customers. This leads to a contrast between the "lit" areas, where information on FX quotes and trades is publically available on some of the electronic trading platforms; and the "dark pools", where information on customer-direct transactions is initially private. A further contrast is between the human and computer elements. At one extreme are arbitrage strategies formalised in computer algorithms and executed by computer algorithms. At the other extreme are trades where humans are in the strategy and execution driving seats. So the notion of a single FX market is more fiction than fact.

Section 3 considers the account given of FX markets by the efficient markets hypothesis (EHM) that came to be the conventional academic wisdom as to how financial markets work. The EMH postulates that prices, such as those in markets for FX, will fluctuate randomly if they incorporate the information and expectations of all market participants [3]. Weak, semi-strong and strong variants of the EMH are distinguished [4] depending on whether the information available involves just the past history of market prices; other publically available information as well; or privately available information as well. If the EMH held in its strong form, opportunities to profit from arbitrage would be fleeting and would not persist in the absence of continuing information surprises [5]. We argue that the EHM belies too many facts for comfort.

Section 4 considers the account of financial markets in the novel *The Fear Index* [6]. In this thriller a particle physicist leaves his research post at the Large Electron-Positron Collider at CERN, Geneva to become a "quant", partnering a banker to set up a hedge fund based on an algorithmic-learning trading strategy. The fictional VIXAL-4 fund makes huge profits by *complex event processing* arbitrage based on algorithms "learning" how fear indices affect asset prices. Aspects of the psychology and neuroscience literature dealing with financial markets are considered in an attempt to identify a factual basis for this novel.

Section 5 considers *classical arbitrage* sequences in FX markets. The starting point is a three-currency world in which one arbitrage transaction suffices to exploit and eliminate FX price discrepancies. The results for a four-currency world are then discussed, the periodicity implications of the combinatorial analysis of arbitrage sequences in asynchronous systems in [1] being highlighted. The implications of moving to a five or higher order currency world [7] are then considered. The *Financial Times* provides daily quotes for fifty-two currencies, so the higher order currency worlds are worthy of consideration.

Section 6 offers some concluding remarks.

2. FX Markets

The Bank for International Settlements conducts a triennial survey that provides useful documentation on global FX market activity [8]. The UK is the dominant location for FX market activity, accounting for 37% of global turnover, followed by the US (18%), EMU countries (10%), Japan (6%), Singapore (5%), Switzerland (5%), Hong Kong (5%) and Australia (4%). There tends to be a "home bias" between the different time zones, most trades initiated in the US and Canada occurring during North American trading hours, most trades initiated in Japan and Australia taking place during Asian trading hours, and UK-initiated trades tending to be bunched in the overlapping Europe-North America and Asia-Europe trading hours [9, Table 2].

FX trades involving the (US) dollar account for 85/200 of all trades, followed by the euro (39/200), yen (19/200) and sterling (13/200), so trades involving these four currencies account for 78% of global turnover [8, p. 12]. In terms of currency pairs the dollar–euro accounts for 28% of global turnover, followed by the dollar–yen (14%), dollar–sterling (9%), euro–yen (3%), euro–sterling (3%) and sterling–yen (0.5%), so these six pairs account for 57.5% of global turnover [8, p. 5]. Triangular arbitrage operations involving these currency pairs are considered in the four-currency case analysed in Section 5 of this paper.

Over the last decade there has been a substantial increase in the proportion of FX transactions that are executed electronically [8, p. 14]. Electronic broking systems, such as the London-based EBS and Reuters, accounted for 18.8% of turnover. Multi-bank electronic communication networks (ECNs) such as the US-based Currenex, Hotspot FX and FXall, accounted for 11.1% of turnover, with single-bank ECNs accounting for 11.4% of global turnover. Most inter-dealer trades are also executed electronically, accounting for 18.5% of turnover. The non-electronic execution part of the market is made up of customer-direct transactions, where banks deal directly with end-user clients, accounting for 24.3% of turnover; and voice broker transactions, which hark back to the method of executing transactions by telephone, which was the predominant method of executing FX transactions until the late 1980s, taking 15.9% of turnover [8, p. 16].

In the 1990s electronic trading was confined to the inter-dealer section of the FX market. The growth of electronic trading in the 2000s has seen the emergence of new trading platforms and new participants in FX markets. The growth of high-frequency trading (HFT) has been one of the most notable developments. Algorithms can be used to execute trades, such as by "splitting trades to minimise the footprint on the market" [2, p. 3]; or to make trading strategy decisions with the algorithm being a *classical arbitrage, latency arbitrage, liquidity imbalance or complex event processing model.*

HFT firms use algorithms for decision-taking as well as execution of FX trades. The latency, that is the delay between the transmission of information from a source and the reception at its destination, of HFT trades can be less than a millisecond, 10^{-3} seconds — compared to around 150 milliseconds for a human blink [2, p. 4]. HFT firms trading FX tend to operate on high volumes and low margins. Access to infrastructure support is by prime brokerage arrangements with large investment banks, as is credit, though risk holding periods are "typically well under five seconds" [2, p. 1]. On March 2011 the EBS introduced a fifth decimal to the pricing of the major currency pairs, this finer grain "pip" being easier to handle for the algorithms than for the human traders. The estimate is that automated HFT accounted for 24-30% of total spot FX turnover in 2010 [2, p. 11].

A pioneering study by Chaboud, Chiquoine, Hjalmarsson and Vega [10] used one-second frequency data from the EBS platform to assess how computer and human-driven FX trades interact, distinguishing between computer-maker/computer-taker, computer-maker/human-taker, human-maker/computer-taker and human-maker/human-taker trades. They found that computer-driven trades tend to be more correlated than human-driven trades, suggesting that the algorithms employed are not very diverse. There was no evident relationship between computer-driven trades and higher exchange-rate volatility, and non-algorithmic order flow accounted for a higher share of the variance in FX trading returns than algorithmic order flow.

3. The EMH and FX Markets

If the EMH holds, foreign exchange rates would fluctuate randomly provided that they fully incorporate the information and expectations of all market participants [3]. To test the EMH, auxiliary hypotheses need to be invoked to specify, inter alia, the information sets available. Fama [4] distinguished three variants of the EMH that could be subjected to statistical tests. In the weak version only information about present and past values of, in this context, exchange rates is involved: so technical analysts or algorithms would not be able to make arbitrage profits by attempting to identify patterns in past exchange rate movements, or discrepancies between exchange rates for currency pairs and their cross-exchange rates. In the semi-strong form the EMH involves publicly available information about the "fundamentals" that determine exchange rates as well: so technical analysts and algorithms would not be able to profit from using macroeconomic news, order flow information available on electronic trading platforms or other relevant driving forces when conducting arbitrage operations. In the strong variant of the EMH, privately available information is also reflected in actual exchange rates: so, for example, FX traders at an investment bank would not be able to make arbitrage profits by exploiting the order-flow information arising from FX trades conducted directly with end-user clients.

Thus, under the weak version of the EMH, exchange rate movements should be random, and arbitrage possibilities arising from discrepancies between the exchange rates for currency pairs and their cross exchange rates would not exist. Under the *semi-strong* and *strong variants* of EMH, technical analysts or computer algorithms would not be able to detect systematic relationships between exchange rates and, respectively, public or public and private information sets. This raises the "arbitrage paradox" pointed out by Grossman and Stiglitz [11]. If arbitrage opportunities do not exist, FX market participants would not have any incentive to monitor exchange rate movements, in which case profitable arbitrage opportunities could well arise. The informal resolution of this paradox by adherents of the EMH is to suggest that arbitrage opportunities arise over short time intervals and are quickly eliminated. Given that the HFT firms can execute trades within a 10^{-3} second interval on electronic FX trading platforms, the short-time interval would have to be very short indeed. A further basic problem with the EMH is the "no trade" theorem [12]. If markets such as those for FX were informationally efficient, one would expect to observe periods in which transactions volumes vanished towards zero when there is no new information coming on stream. Instead FX transactions, although varying over time, remain substantial during quiescent market conditions.

An influential paper by Meese and Rogoff [13] produced evidence, for low-frequency exchange rate data, that models in which macroeconomic "fundamentals" drive exchange rates did not forecast any better than the EMH postulate that the changes are random. Engel and West [14] showed that exchange rates would display something close to a random walk if the "fundamentals" followed an I(1) process and the factor for discounting future fundamentals was close to one.

The microstructure literature on the FX market has focused attention on the role of order-flow information in explaining exchange rate changes. Some of this information is private, arising, for example, from FX transactions in which banks deal directly with end-user clients. Evans and Lyons [15] showed that end-user order flow data can explain about 16% of the monthly spot rate variance between the dollar and the euro. Covrig and Melvin [16] studied whether geographically local order flow conveys informational advantages. Until December 22, 1994 the Japanese FX market closed for lunch from 12.00 to 13.30 hours, Tokyo time. The authors found bunching of FX transactions before this lunchtime closure, and a significant tendency for foreign quotes for the yen–dollar exchange rate to lag behind Tokyo quotes during this pre-lunch period. This suggests that Tokyo based traders were better informed, or that foreign-based traders believed this to be the case. Osler [17] discovered significant effects of stop-loss order flows on minute-by-minute foreign exchange rate data, a stop-loss order being an instruction to buy(sell) at the FX platform's spot rate once this has risen (fallen) to a pre-specified level.

Studies of high frequency data on electronic trading platforms suggest that arbitrage trading profits could have been made, net of transactions costs. Data from the EBS platform were studied by Marshall, Treepongkaruna and Young [18] to estimate triangular arbitrage opportunities between the dollar, euro, yen, sterling and Swiss franc during two-minute intervals. Estimated mean arbitrage profits, net of bid-offer spreads and 0.2 basic point transaction fees, ranged from 2.8 to 3 basis points. Reuters platform data were used in [19] to detect covered interest arbitrage (CIP) opportunities, where the CIP no-arbitrage condition is that the forward-spot exchange rate differential matches the interest rate differentials between similar assets with the same maturity. After taking account of transaction costs, the authors discovered numerous short-lived deviations from the CIP condition that would have presented profitable arbitrage opportunities, and that the profit opportunities were large enough and of significantly long duration for traders to have exploited them with profit.

4. The Fear Index

If the financial markets are informationally efficient, as the EMH postulates, it would not be possible to design investment strategies that would consistently outperform the market. In the EMH world it is difficult to explain the excess returns of the funds managed by the likes of Soros [20], whose early investment ventures involved arbitrage between the US and European markets, and Buffett [21]. Such funds have not always outperformed indexed portfolios, but have shown a persistent tendency to do so.

Samuelson, the co-founder of the EMH, pointed out that the EMH implies that "most portfolio decision makers should go out of business — take up plumbing, teach Greek, or help produce the annual GDP by serving as corporate executives... even if this advice to drop dead is good advice, it is obviously not counsel that will be eagerly followed" (1974, quoted in [22, p. 113]. Samuelson himself did not follow the EMH counsel when, in 1970 he became a founder investor in Commodities Corporation, started by a Ph.D. graduate from Samuelson's MIT. This fund used statistical models of supply and demand for commodities, collecting detailed information on weather conditions affecting cocoa harvests, for example, to forecast commodity prices [23, Ch. 3]. As Samuelson later explained: "Fama's theory of the random walk and mine are not the same... mine is that there are no easy pickings... if you read the numerous papers I have written on the efficient markets hypothesis, you will realise it is not a dogma... if you get information early, before it is widespread, you can't help but get very rich" (Samuelson, February 5, 2008, interview quoted in [23, p. 418].

If the EMH were true, and there were no persistent arbitrage profits to be made in financial markets such as those for FX, it is difficult to explain the existence of "quants", mathematicians and physicists employed by financial institutions to detect patterns in the processes determining financial asset prices. The "quants" have formalised in algorithms the trends or tendencies that "chartists" tried to identify before the advent of powerful computers. A battalion of "quants" form the workforce in the hedge fund described in the novel *The Fear Index* [6]. This fund, *Hoffmann Investment Technologies*, has been set up by Alex Hoffmann, funded by a partnership with a former investment banker, Hugo Quarry. Hoffmann has a Ph.D. in physics from Princeton, and was employed on the Large Electron-Positron Collider project, the forerunner to the Large Hadron Collider, at CERN in Geneva, leaving under a cloud.

The VIXAL-4 fund operated by *Hoffmann Investment Technologies* has an investment strategy based on algorithms that identify patterns in the determination of asset prices, and also on algorithms to execute trades. A first innovation in the VIXAL-4 strategy is to "correlate recent market fluctuations with the frequency rate of fear-related words in the media — terror, alarm, panic, horror, dismay, dread, scare, anthrax, nuclear" [6, p. 97]. Trades are geared to the volatility index VIX, a measure of volatility in the Standard and Poor 500 index of US equity prices, tradable on the Chicago Board of Exchange. The second innovation is to allow the algorithmic trading strategy to evolve as new data emerges, as the market impact of the trades made is observed, and as the profit and losses associated with particular trades are revealed. Hence the evolution from VIXAL-1 to VIXAL-4 in the machine-learning algorithmic trading strategy. There is a Darwinian theme, the innovative use of fear indices being depicted through the lens of Darwin's *The Expression of the Emotions in Man and Animals* [24]; and the use of evolutionary adaptation reflecting Darwin's *On the Origin of the Species by Means of Natural Selection* [25].

Reading a novel raises the problem of distinguishing the factual basis from the fiction. In relation to FX arbitrage processes an intriguing question is whether the trading processes described in The Fear Index give a closer insight into what happens in actual FX markets than does the vision of this world provided by the EMH. In *The Fear Index* emotions play a key role in driving behaviour on financial markets, whereas in the EMH behaviour is driven by reason alone. In the novel the trading strategy algorithm adapts to changes in the market

Journal of Physics: Conference Series **585** (2015) 012015 doi:10.1088/1742-6596/585/1/012015

environment, whereas in the EMH there is no incentive to search for fitter trading strategies.

The notion of rational economic man that underlies the description of financial markets provided by the EMH is difficult to square with much of the language used to describe such markets: market crashes, panic selling, herding, bull markets, bear markets, fear, market sentiment, optimism, pessimism and so on. The psychology literature is replete with examples of cognitive biases in the heuristics used by people when asked to trade on experimental markets, such as risk aversion, anchoring, framing, representativeness and over-confidence [26]. The neuroscience literature suggests that the amygdala, a part of the brain that plays a key role when there is a need to take account of both affective and cognitive responses when making decisions, is highly active when financial trading decisions are being made [27]. Sweat samples taken from traders on a London trading floor indicated that higher testosterone levels predicted greater profitability later in the trading day, and that cortisol levels were positively correlated with market volatility [28]. When financial trading can be driven by emotions, the question arises of how the traders react to the profits or losses made. The adaptive markets hypothesis [29] postulates that trading strategies evolve, with the "fitter" ones showing higher survival propensity. This is not too far from the trading world depicted in the early stages of *The Fear Index* — before the VIXAL-4 fund finds its Nemesis.

5. FX Trading Sequences

In what follows we consider a particular type of *classical arbitrage* strategy that operates on discrepancies between the exchange rates for currency pairs and their cross-exchange rates. This means that we do not consider other *classical arbitrage* strategies, such as that based on deviations from the covered interest parity (CIP) condition. Nor do we consider *latency, liquidity imbalance* or *complex event processing* arbitrage strategies. The focus is on the different combinations of arbitrage sequences that can be pursued, and on whether the pursuit of arbitrage transactions that are active, in the sense that they are profitable, leads to a no-arbitrage state in which there are no active arbitrage opportunities. The following exposition is based on [1].

5.1. Three Currencies

In a three-currency world arbitrage operations are reasonably straightforward. Let the currencies be the US dollar ($), euro (€) and sterling (£), and their exchange rates (r) be defined as the number of units of domestic currency required to buy (or sell — we ignore the bid-ask spread between exchange rates and transaction costs in what follows) one unit of foreign currency. So, for example, $r_{\$€}$ is the number of dollars required to buy one euro. Thus, ignoring reciprocal exchange rates, the principal exchange rates are the triplet:

$$(r_{\$€},\ r_{\$£},\ r_{€£})$$

Let there be three currency traders who, in line with the "home currency bias" evidence mentioned earlier in this paper, initially are aware only of the exchange rates involving their home currency. So, the dollar trader does not initially know the value of $r_{€£}$, the euro trader is initially unaware of $r_{\$£}$ and the sterling trader is initially unaware of $r_{\$€}$.

The condition for the dollar trader to conduct profitable arbitrage, when discovering that the euro–sterling exchange rate is out of line with the dollar exchange rate, is:

$$r_{\$£} \cdot r_{£€} > r_{\$€}.$$

The dollar trader would then make an arbitrage profit by exchanging dollars for pounds and using the pounds to buy euros. This would continue until the new non-arbitrage exchange rate is reached:

$$r_{\$€}^{(new)} = \frac{r_{\$£}}{r_{€£}},\quad r_{\$£},\quad r_{€£}.$$

Journal of Physics: Conference Series **585** (2015) 012015 doi:10.1088/1742-6596/585/1/012015

The euro trader would have an active arbitrage opportunity if he discovered that the dollar–sterling rate was such that:

$$r_{€\$} \cdot r_{\$£} > r_{€£}.$$

Here the euro trader would profit by exchanging euros for dollars and the dollars for sterling until a new no-arbitrage ensemble of exchange rate is established:

$$r_{\$€}, \quad r_{\$£}, \quad r_{€£}^{(new)} = \frac{r_{\$£}}{r_{\$€}}.$$

Finally, the sterling trader could discover that the dollar–euro exchange rate is such that:

$$r_{£€} \cdot r_{€\$} > r_{£\$}.$$

Here the sterling trader would profit by exchanging sterling for euros and the euros for dollars until a new no-arbitrage ensemble emerges:

$$r_{\$€}, \quad r_{\$£}^{(new)} = r_{\$€} \cdot r_{€£}, \quad r_{€£}.$$

In this three-currency world the order in which traders find information about active arbitrage opportunities matters, but one arbitrage operation, perhaps executed algorithmically to minimise market impact, would suffice to exploit and, by doing so, remove the profitable arbitrage opportunity. This is not too far from the EMH world in which active arbitrage opportunities are, at most, fleeting. The interesting question is whether this state of affairs pertains when a larger number of currencies are present.

5.2. Four Currencies

Let the fourth currency be the Japanese yen (¥). Then we have $C_4^2 = 6$ principal exchange rates with $\mathcal{R}$ indicating this ensemble:

$$\mathcal{R} = (r_{\$€}, \ r_{\$£}, \ r_{\$¥}, \ r_{€£}, \ r_{€¥}, \ r_{£¥})$$

The balanced set of exchange rates, in which no arbitrage opportunities exist, can be shown by inspection, to be:

$$r_{€£} = \frac{r_{\$£}}{r_{\$€}}, \quad r_{€¥} = \frac{r_{\$¥}}{r_{\$€}}, \quad r_{£¥} = \frac{r_{\$¥}}{r_{\$£}}.$$

Again, it is assumed that each trader is initially aware only of the exchange rates for his home currency. So, for example, the dollar trader knows $r_{\$¥}$, $r_{\$£}$ and $r_{\$¥}$ but not $r_{€£}$, $r_{€¥}$ and $r_{£¥}$.

The crucial issue in this four-currency world is the order in which traders discover information about profitable arbitrage opportunities. So, for example, if the first information about a cross-exchange rate arbitrage opportunity arrives in the form of the dollar trader finding out that the euro–sterling exchange rate is out of line with the rates for the dollar–euro and dollar–sterling currency pairs, then a dollar–euro–sterling ($€£) arbitrage sequence will be exploited first. After the exploitation of this triangular arbitrage opportunity, which is denoted as $\mathcal{A}_{\$€£}$, a new ensemble of principal exchange rates would emerge:

$$\mathcal{R}^{(new)} = \mathcal{R}\mathcal{A}_{\$€£} = \left(r_{\$€}^{(new)}, \ r_{\$£}, \ r_{\$¥}, \ r_{€£}, \ r_{€¥}, \ r_{£¥}\right).$$

Where $\mathcal{R}\mathcal{A}_{\$€£}$ indicates the new ensemble of exchange rates after an active arbitrage $\mathcal{A}_{\$€£}$, has been exploited. This new exchange rate ensemble would present new cross-exchange rate arbitrage opportunities to other traders. So the crucial difference between the three and four-currency case is that in the latter case a single arbitrage opportunity does not bring the FX

Journal of Physics: Conference Series **585** (2015) 012015 doi:10.1088/1742-6596/585/1/012015

market into a balanced ensemble of exchange rates in which no active arbitrage opportunities exist. The exploitation of one arbitrage opportunity has a "ripple" effect, the change in cross-exchange rates creating further arbitrage opportunities about which the other traders could become informed.

The number of possible triangular arbitrage operations possible in this four-currency case is ${}^3P_4 = 24$. These possible arbitrage operations are listed in Table 1 (source: [1]), along with their activation conditions and the new exchange rates arising from their exploitation:

Table 1. List of arbitrages

Number	Arbitrage	Activation condition	Actions
1	$\mathcal{A}_{\$€£}$	$r_{€£} > r_{\$€} \cdot r_{\$£}$	$r^{(new)}_{\$€} = r_{\$£}/r_{€£}$
2	$\mathcal{A}_{\$€¥}$	$r_{\$¥} > r_{\$€} \cdot r_{€¥}$	$r^{(new)}_{\$€} = r_{\$¥}/r_{€¥}$
3	$\mathcal{A}_{\$£€}$	$r_{\$€} \cdot r_{€£} > r_{\$£}$	$r^{(new)}_{\$£} = r_{\$€} \cdot r_{€£}$
4	$\mathcal{A}_{\$£¥}$	$r_{\$¥} > r_{\$£} \cdot r_{£¥}$	$r^{(new)}_{\$£} = r_{\$¥}/r_{£¥}$
5	$\mathcal{A}_{\$¥€}$	$r_{\$€} \cdot r_{€¥} > r_{\$¥}$	$r^{(new)}_{\$¥} = r_{\$€} \cdot r_{€¥}$
6	$\mathcal{A}_{\$¥£}$	$r_{\$£} \cdot r_{£¥} > r_{\$¥}$	$r^{(new)}_{\$¥} = r_{\$£} \cdot r_{£¥}$
7	$\mathcal{A}_{€\$£}$	$r_{\$£} < r_{\$€} \cdot r_{€£}$	$r^{(new)}_{\$€} = r_{\$£}/r_{€£}$
8	$\mathcal{A}_{€\$¥}$	$r_{\$¥} < r_{\$€} \cdot r_{€¥}$	$r^{(new)}_{\$€} = r_{\$¥}/r_{€¥}$
9	$\mathcal{A}_{€£\$}$	$r_{\$£} > r_{€£} \cdot r_{\$€}$	$r^{(new)}_{€£} = r_{\$£}/r_{\$€}$
10	$\mathcal{A}_{€£¥}$	$r_{€¥} > r_{€£} \cdot r_{£¥}$	$r^{(new)}_{€£} = r_{€¥}/r_{£¥}$
11	$\mathcal{A}_{€¥\$}$	$r_{\$¥} > r_{€¥} \cdot r_{\$€}$	$r^{(new)}_{€¥} = r_{\$¥}/r_{\$€}$
12	$\mathcal{A}_{€¥£}$	$r_{€£} \cdot r_{£¥} > r_{€¥}$	$r^{(new)}_{€¥} = r_{€£} \cdot r_{£¥}$
13	$\mathcal{A}_{£\$€}$	$r_{\$€} \cdot r_{€£} < r_{\$£}$	$r^{(new)}_{\$£} = r_{\$€} \cdot r_{€£}$
14	$\mathcal{A}_{£\$¥}$	$r_{\$¥} < r_{\$£} \cdot r_{£¥}$	$r^{(new)}_{\$£} = r_{\$¥}/r_{£¥}$
15	$\mathcal{A}_{£€\$}$	$r_{\$£} < r_{€£} \cdot r_{\$€}$	$r^{(new)}_{€£} = r_{\$£}/r_{\$€}$
16	$\mathcal{A}_{£€¥}$	$r_{€¥} < r_{€£} \cdot r_{£¥}$	$r^{(new)}_{€£} = r_{€¥}/r_{£¥}$
17	$\mathcal{A}_{£¥\$}$	$r_{\$¥} > r_{£¥} \cdot r_{\$£}$	$r^{(new)}_{£¥} = r_{\$¥}/r_{\$£}$
18	$\mathcal{A}_{£¥€}$	$r_{€¥} > r_{£¥} \cdot r_{€£}$	$r^{(new)}_{£¥} = r_{€¥}/r_{€£}$
19	$\mathcal{A}_{¥\$€}$	$r_{\$€} \cdot r_{€¥} < r_{\$¥}$	$r^{(new)}_{\$¥} = r_{\$€} \cdot r_{€¥}$
20	$\mathcal{A}_{¥\$£}$	$r_{\$£} \cdot r_{£¥} < r_{\$¥}$	$r^{(new)}_{\$¥} = r_{\$£} \cdot r_{£¥}$
21	$\mathcal{A}_{¥€\$}$	$r_{\$¥} < r_{€¥} \cdot r_{\$€}$	$r^{(new)}_{€¥} = r_{\$¥}/r_{\$€}$
22	$\mathcal{A}_{¥€£}$	$r_{€£} \cdot r_{£¥} < r_{€¥}$	$r^{(new)}_{€¥} = r_{€£} \cdot r_{£¥}$
23	$\mathcal{A}_{¥£\$}$	$r_{\$¥} < r_{£¥} \cdot r_{\$£}$	$r^{(new)}_{£¥} = r_{\$¥}/r_{\$£}$
24	$\mathcal{A}_{¥£€}$	$r_{€¥} < r_{£¥} \cdot r_{€£}$	$r^{(new)}_{£¥} = r_{€¥}/r_{€£}$

Journal of Physics: Conference Series **585** (2015) 012015 doi:10.1088/1742-6596/585/1/012015

The key question, then, is whether the "ripple" effects arising from the exploitation of an active arbitrage opportunity results in a sequence of active arbitrages that converge onto a no-arbitrage ensemble of exchange rates. Cross et. al. [1] apply asynchronous systems analysis to investigate the way active arbitrage chains evolve. The key finding is that "rather than there being a smooth convergence to an ensemble of exchange rates with no arbitrage opportunities, the arbitrage operations may display periodicity and no necessary convergence on a cross-exchange rate law of one price" [1, p. 252]. This periodicity finding is interesting in several respects. It suggests that active arbitrage opportunities will continue to exist in markets such as those for FX. Rather than the exploitation of active arbitrage opportunities leading to their elimination, their exploitation is followed by "ripple " effects that do not necessarily tend to fade away, new arbitrage opportunities being activated.

Preliminary results [7] suggest that triangular arbitrage in a world of five or more currencies and traders involves more unstable behaviour in arbitrage chains than in the four-currency case. In this higher currency-dimension world, exchange rates not only may display the periodicity arising in the four-currency case, "but also, what is surprising, may grow in accordance with the double exponential law" [7, p. 11]. For instance, in the five-currency case exchange rates may grow as $e^{c \cdot 1.071^n}$, where n is the number of arbitrage operations and c is a positive constant. When the number of currencies and currency-traders involved is greater than five then we face an effect of "inheritance of instability" taking place in the cases of four and five currencies. In this case one may observe all the types of behaviour of exchange rates described above: periodicity, growth with the exponential rate and growth in accordance with the double exponential law.

6. Concluding Remarks

The EMH postulates that financial markets are informationally efficient, so that price movements are essentially random. Applied to FX markets, the EMH implies that profitable arbitrage opportunities are, at most, short-lived. This paper has documented the fragmented nature of FX markets, the institutional fragmentation implying that the information about profitable arbitrage opportunities available to FX participants is also fragmented. Even on single electronic trading platforms there is evidence from historical data that triangular arbitrage operations would have been profitable, net of bid-ask spreads and transaction costs.

In the world described by the EMH it is difficult to explain the existence of funds, such as the hedge fund in *The Fear Index* [6], that employ "quants" to design algorithms to detect arbitrage opportunities. The problem here might be that the academic literature has failed to analyse the complexity of the combinations of possible arbitrage sequences in markets such as those for FX. The combinatorial analysis of such sequences is straightforward in a three-currency world. As shown in [1], in a four-currency world profitable arbitrage sequences tend to display periodicity. This finding helps to explain why arbitrage operations in markets such as those for FX are endemic rather than self-extinguishing.

References

[1] Cross R, Kozyakin V, O'Callaghan B, Pokrovskii A and Pokrovskiy A 2012 *Metroeconomica* **63** 250–294

[2] BIS 2011 *High-Frequency Trading in the Foreign Exchange Market: New Report from the Markets Committee* (Basel, Switzerland: Bank of International Settlement)

[3] Samuelson P 1965 *Industrial Management Review* **6** 41–49

[4] Fama E F 1970 *Journal of Finance* **25** 383–417 Papers and Proceedings of the Twenty-Eighth Annual Meeting of the American Finance Association New York, N.Y. December, 28–30, 1969 (May, 1970)

[5] Fisher F M 1989 *Disequilibrium Foundations of Equilibrium Economics* Econometric Society Monographs (Cambridge: Cambridge University Press)

[6] Harris R 2011 *The Fear Index* (London: Hutchinson)

[7] Cross R and Kozyakin V 2012 *ArXiv.org e-Print archive* (*Preprint* 1204.3422)

[8] BIS 2010 *Triennial Central Bank Survey of Foreign Exchange and Derivatives Market Activity in 2010 — Final Results* (Basel, Switzerland: Bank of International Settlement)

[9] D'Souza C 2008 *Bank of Canada Review* **Spring 2008** 19–27
[10] Chaboud A, Chiquoine B, Hjalmarsson E and Vega C 2009 Rise of the machines: Algorithmic trading in the foreign exchange market International Finance Discussion Papers 980 Board of Governors of the Federal Reserve System (U.S.)
[11] Grossman S J and Stiglitz J E 1976 *American Economic Review* **66** 246–253
[12] Grossman S J and Stiglitz J E 1980 *American Economic Review* **70** 393–408
[13] Meese R A and Rogoff K 1983 *Journal of International Economics* **14** 3–24
[14] Engel C and West K D 2005 *Journal of Political Economy* **113** 485–517
[15] Evans M D D and Lyons R K 2005 *American Economic Review* **95** 405–414
[16] Covrig V and Melvin M 2002 *Journal of Empirical Finance* **9** 271–285
[17] Osler C L 2005 *Journal of International Money and Finance* **24** 219–241
[18] Marshall B R, Treepongkaruna S and Young M 2007 Exploitable arbitrage opportunities exist in the foreign exchange market Discussion Paper, 10 September, Massey University, Palmerston North, New Zealand
[19] Akram Q F, Rime D and Sarno L 2008 *Journal of International Economics* **76** 237–253
[20] Soros G 1995 *Soros on Soros: Staying Ahead of the Curve* (John Wiley & Sons Inc)
[21] Lowenstein R 2008 *Buffett: The Making of an American Capitalist* (New York: Random House)
[22] Bernstein P L 2007 *Capital Ideas Evolving* (Hoboken, New Jersey, USA: John Wiley and Sons)
[23] Mallaby S 2010 *More Money Than God: Hedge Funds and the Making of the New Elite* (London, UK: Bloomsbury Publishing)
[24] Darwin C 1872 *Expression of the Emotions in Man and Animals* (London, UK: John Murray)
[25] Darwin C 1859 *On the Origin of Species by Means of Natural Selection* (London, UK: John Murray)
[26] Kahenman D 2011 *Thinking, Fast and Slow* (London, UK: Allen Lane)
[27] Lo A 2011 Fear, greed and financial crises: a cognitive neurosciences perspective Tech. rep. MIT Sloan School of Management Boston, MA, USA
[28] Coates J M and Herbert J 2008 *Proceedings of the National Academy of Sciences USA* **105** 6167–6172
[29] Lo A 2004 *Journal of Portfolio Management* **30** 15–29